One God, Two Goddesses, Three Studies of South Indian Cosmology

Jerusalem Studies in Religion and Culture

Editors

VOLUME 18

The titles published in this series are listed at brill.com/jsrc

One God, Two Goddesses, Three Studies of South Indian Cosmology

by

Don Handelman

BRILL

LEIDEN • BOSTON
2014

Library of Congress Cataloging-in-Publication Data

Handelman, Don.
One god, two goddesses, three studies of South Indian cosmology / by Don Handelman.
pages cm. — (Jerusalem studies in religion and culture, ISSN 1570-078X ; volume 18)
Includes bibliographical references and index.
ISBN 978-90-04-25615-6 (hardback : alk. paper) — ISBN 978-90-04-25739-9 (e-book : alk. paper)
1. Hinduism—India, South—Rituals. 2. Murugan (Hindu deity)—Cult. 3. Hindu goddesses—India, South. 4. Hindu cosmology. I. Title.

BL1226.2.H36 2013
294.5'211—dc23

2013026797

This publication has been typeset in the multilingual "Brill" typeface. With over 5,100 characters covering Latin, IPA, Greek, and Cyrillic, this typeface is especially suitable for use in the humanities. For more information, please see www.brill.com/brill-typeface.

ISSN 1570-078x
ISBN 978-90-04-25615-6 (hardback)
ISBN 978-90-04-25739-9 (e-book)

This book is printed on acid-free paper.

FSC www.fsc.org MIX Paper from responsible sources FSC® C109576

Printed by Printforce, the Netherlands

In memory of the late Galina Lindquist
who had the gift of true sight

CONTENTS

LIST OF ILLUSTRATIONS

ACKNOWLEDGMENTS

I wish to thank the many people who helped us, in many ways, in the course of our work in Vizianagaram: the late Bantupalli Bairagi Nayudu, above all; Srimati Bantupalli Varalakshmi; Bantupalli Venkata Rao and Bantupalli Bala Krishna; Srimati Netati Vijaya Nirmala and Sri Srinivas; Srimati Netati Sobha Rani and Sri Sankara Rao; Sri Ramavarapu Paidiraju and his troupe; Sri Tallapudi Bhaksara Rao (the Sirimanu pujari at present); Sri Maipilli Bhaskara Satyanarayana; Members of Sri Paidimamba Swaschanda Seva Snagham in Hukhumpeta, Vizianagaram; Raja Saheb Dr. Pusupati Ananda Gajapati Raju and his wife, Srimati Pusupati Sudha Gajapati Raju and their daughter, Kum. Urmila Gajapati Raju; Raja Saheb Sri Pusupati Ashoka Gajapati Raju; Dr. Devadula Ramalingeswara Swamy, an Ayurveda Vaidyacharya. Without the continuous support and detailed conversations with all of these friends, the Paiditalli essay could never have been completed.

Four students accompanied us to Vizianagaram: Dr. Nir Avieli; Aya Tamal; Neta Bar; and Dr. Boaz Amichay. Their insightful reactions to what we saw there helped shape our understandings of the goddess.

I also wish to thank the late R. V. G. K. Ranga Rao, Raja of Bobbili, for his gracious hospitality, and R. V. S. K. Ranga Rao, the present Raja, along with R. V. R. K. Ranga Rao, V. S. K. K. Ranga Rao, and Chodiganji Ramesh Naidu. For much help and good cheer, I thank Prof. U. Tataji, formerly Vice-Chairman, Andhra Pradesh State Council for Higher Education and Emeritus Professor of Sociology, Andhra University. At Tirupati our work was put on track largely through the direction of Dr. A. Anand.

My thanks to David Shulman, Velcheru Narayana Rao and Joyce Burckhalter Flueckiger together with whom I did the fieldwork in Tirupati. David was the spark of this research, Narayana Rao the wellspring, and Joyce, most of all, helped me to perceive the goddess as a warm, vibrant, caring woman.

M. V. Krishnayya opened Vizianagaram to David and myself. A professor of western philosophy at Andhra University, Krishnayya is no less a devoted ethnographer of Andhra folkways, documenting numerous sites of ritual with undying curiosity and insight. Krishnayya's engaging and sharing presence often made the difference.

Chapter Two is reprinted with the kind permission of University of Chicago Press from *History of Religions* 27:2 (1987), pp. 133–170.

Chapter Three is reprinted from *Syllables of Sky: Studies in South Indian Civilization in honour of Velcheru Narayana Rao*, edited by D. Shulman, Oxford University Press, Delhi, 1995, pp. 283–337.

CONTRIBUTORS

Don Handelman is Sarah Allen Shaine Professor Emeritus of Anthropology at the Hebrew University, Jerusalem. He has published extensively on ritual and religion, including *Models and Mirrors: Towards an Anthropology of Public Events* (2nd ed. Berghahn, 1998) and *Nationalism and the Israeli State: Bureaucratic Logic in Public Events* (Berg, 2004). David Shulman and he have essayed southern Indian cosmologies in, *God Inside Out: Śiva's Game of Dice* (Oxford, 1997) and *Śiva in the Forest of Pines: An Essay on Sorcery and Self-Knowledge* (Oxford, 2004). Together with Galina Lindquist he has edited, *Ritual in its Own Right: Exploring the Dynamics of Transformation* (Berghahn, 2005) and *Religion, Politics, and Globalization: Anthropological Approaches* (Berghahn, 2011). With T. M. S. Evens he has edited, *The Manchester School: Practice and Ethnographic Praxis in Anthropology* (Berghahn, 2006).

Bruce Kapferer is Professor of Social Anthropology at the University of Bergen and Honorary Professor of Anthropology at University College London. He has published monographs based on his fieldwork in Central Africa, India and Sri Lanka and Australia. Much of his research has concentrated on healing rites in Sri Lanka and more recently on teyyam performances in Kerala. Major books include *A Celebration of Demons* (Indiana University Press, 1983), *The Feast of the Sorcerer* (University of Chicago, 1997), *Legends of People, Myths of State* (Smithsonian Institution Press, 1988, New Edition and Berghahn, 2011). He is currently Editor of Social Analysis (Berghahn) and the journal of Anthropological Theory (Sage).

M. V. Krishnayya is Professor Emeritus of Philosophy at Andhra University, Visakhapatnam and the leading scholar of village rituals and local deities in the coastal districts of Andhra Pradesh, from Srikakulam to Rajahmundry. Among his recent publications: "One Who Stays for Good: Perantallu in Northern Coastal Andhra," Indian Folklore Research Journal (2010). Trained as a philosopher specializing in existentialism, Advaita, and Gandhian theory and praxis, he has devoted much of his work to long-term anthropological fieldwork in the Telugu-speaking region.

David Shulman is Renee Lang Professor of Humanistic Studies at the Hebrew University, Jerusalem. A philologist specializing in languages and literatures of southern India, particularly Tamil, Telugu, Malayalam, and Sanskrit, he has been tempted into fieldwork with Don Handelman, M. V. Krishnayya, and Joyce B. Flueckiger. He has published widely on the cultural and religious history of southern India and worked closely with Velcheru Narayana Rao, Sanjay Subrahmanyam, and Don Handelman. Among his recent works is *More than Real: A History of the Imagination in South India* (Harvard University Press, 2012). His deeper interest lies in Carnatic music.

PREFACE

When I was young and in the throes of shaping an anthropology for myself I never thought of South Asia as a desirable locale for research. My fieldwork moved from the Great Basin to Israel (where I married and lived), then to Newfoundland and back again to Israel. In 1979 Bruce Kapferer invited Lea, my late wife, and me to spend time with him in Sri Lanka where he was doing fieldwork. Bruce and I were close friends from our Manchester days, and then as now I appreciate his analytical brilliance, his provocative and synthesizing acumen, the power of his intellectual grasp, and his willingness to take on and work through any intellectual challenge. I value serendipity. And that which Jung called synchronicity. My best moments and relationships arrive that way, as unannounced, quiet presences that sometimes are life-changing. My anthropology then and now is to grab onto a strange line of flight and then to hold on for dear life. To wherever. We went and were fascinated. Bruce took us through many byways of Sri Lanka and brought me to see, smell, touch, taste, a vibrant corner of South Asia. He had done me a wonderful turn.

Back in Jerusalem in 1980 a friend told me that I had to meet David Shulman, who had collected and studied Tamil temple myths for his doctorate at SOAS. David was amazing then as he is now, open to everything under the heavens (including the heavens themselves). I was amused to discover that he lived a five-minute walk from my home. David was always supportive, insightful, and encouraging as I slowly found my way through the unfamiliar writings on South Asia, keeping me on track when unwittingly my fantasies took me beyond the possible. David, a poet of the imaginary before he is anything else, is a treasury of knowledge, wisdom, and perception, a polymath by any other name, one who likely thinks that boundaries between academic disciplines are for those who like or need such barriers. Every tyro dreams of such an illuminating guide. The two books we wrote together were a pleasure to do. And every time I write 'then as now' I wonder anew at my good fortune in having friends like these.

Reading on South Asia, doing fieldwork in Andhra Pradesh, was a strange enjoyment. India fills the senses with imaginings, yet these are imaginings within imaginings, fractal imaginings that are borderless and, for me at least, that curve mind-work inward, involuting, yet involution that is emergent, always re-emerging elsewhere into another angle of an expanding cosmos to which I had not had access before. There follow now three moments of emergent imagining that felt cosmic within me, one from each of the locations in the essays of this book.

During the annual Vel Festival the god, Murugan, beloved of both Tamils and Sinhalese, is processed through the thoroughfares of Columbo together with his two wives. Watching this in 1979, bewildered at the bright colors in the steamy heat, the scents of spice and garbage, the enthusiastic devotions of worshippers, the deities being rocked to and fro by a priest as their carriage moves along, I suddenly have a sense of the carriage tracing the form of the cosmic syllable, OM, through the city streets, actualizing the cosmos, sacralizing the city. A moment of cosmic re-creation. My fantasy, of course. Wasn't it?

Toward the end of the searing hot season, the goddess Gangamma annually returns to the small city of Tirupati. There, through ritual, she rejuvenates her cosmos by turning it entirely female, ridding it of all males. Near the end of the visit her head is built on a huge scale outside her temple. This perhaps is the clearest evidence of her totalizing the cosmos she is rejuvenating. On the last day, close to midnight, Gangamma enters her temple courtyard, facing herself facing herself, facing her transformation of cosmos that is fully herself, facing the entirety of cosmos-as-female. Gangamma reaches up and rends her own enormous cheek, and with that the assembled crowd of worshippers surges forward and tears Gangamma to shreds as Gangamma slips away and disappears until the hot season of the following year. In that moment I am struck by just how beautifully calibrated ritual design can be, as the creator and renewer of cosmos destroys her own presence on the hard soil of humankind, once her labors here are fulfilled.

The dwellers in the lands around the small city of Vizianagaram know their country as that of the Mahabharata, its events mapped onto the topography of the land. One climbs the thousand steps of a long hill to the tiny Krsna shrine at the top to where the Pandavas had ascended to ask him to grant them a thirteenth year in exile. Yet unlike us the Pandavas had ascended through the caves and tunnels which fill the mountain with depth. A few miles away next to the shrine of an ancient Siva lingam that thrusts outward from the deepest depths of the earth one finds the tree in whose branches the Pandavas hid their weapons before going into exile. A land of intricate depths.

One twilight I was relaxing on the balcony of a small hotel looking out at the waters of Big Lake, within which the goddess, Paiditalli, had been born, the Old City of Vizianagaram on the far side. The liquid depths of the waters. Porous mountains reflecting in the waters. The conjuncture of so much transformation and continuation in the lengthy association of Paiditalli with Vizianagaram, within which she emerges annually through her own interior fluidity, from her own liquid depths that are her cosmos, in order to grow anew the fruitfulness and vitality of the city. And I felt, indeed felt, an inkling, a momentary shadowy glimpse of just how recursive this cosmos is. Of how the depth of a mountain fits into the depth of a lake, while the porous interior of a mountain (with its swirling caves and twisting tunnels) can take in the sea. Of how in a plowed field the space between one furrow and another is a high mountain ridge, while the furrow itself filled with water is a deep lake. Of how, if the spheroid cosmos is turned on its head, the waters of the lake fall on the land like rain; as rainwater flows down the mountains into the fields and their furrows. Of how all of these surfaces that are depths fit fluidly into one another, and of how this fullness of cosmos becomes immanent as Paiditalli appears in the human world. I went to sleep feeling deepened.

Grab a line of flight not your own and hang on for dear life...

Jerusalem April 2013

CHAPTER ONE

THE COSMIC VORTEX AND DON HANDELMAN'S ANTHROPOLOGICAL VISION: A PERSONAL VIEW

Bruce Kapferer

The three essays presented here are fine examples of a method and a craft of anthropological analysis combined with a conceptual/theoretical vision that is ultimately comparative in its sweep. The idea of craft instead of method communicates for me Don Handelman's distinctive style. Like a master craftsman, Handelman works with his materials. He follows their grain, enters within their spirit, establishing a unity with them and giving expression to their particular dynamic. Handelman combines a highly intuitive feeling and sensitivity for the ethnographic materials that he works with, with a thoroughgoing concern to rigorously establish the analytical and objective force of his arguments and discoveries. He continually opens to new vistas of possibility extending well-beyond conventional or established understandings.

Handelman has maintained a deep fascination with religious, ritual and symbolic materials from the Indian sub-continent for over thirty years.[1] His interest, I think, was sparked during a short trip he made to Sri Lanka in 1979 when he visited the major shrine to Kataragama/Murugan in the island's South Eastern Province, followed up during the same trip with a study of the Vel festival for Murugan in the island's capital of Colombo. This interest was to blossom during a long period of intellectual and field collaboration with David Shulman which resulted in two major co-authored books, *God Inside Out* (1997) and *Śiva in the Forest of Pines* (2004). These two volumes are an exploration into the poetics of central mythological events in Sanskrit, Telugu and literature. Although highly

[1] One should not overlook the longer-term interest that Handelman has shown in ritual practice from very early in his career (see Handelman 1967, 1983). This has also been closely connected to his interest in what has been described as processes of "ritualization," which bear a relation to Handelman's (2004) major interest in "bureaucratic logic" and were developed to the greatest extent in his study of nationalism. Handelman (e.g. 1981, 1996, 2001) has also taken a keen interest in the dynamics of humour and play; these are crucial to his understanding of ritual and, of course, to the kind of analysis that he pursues in these essays.

significant for the understanding of ritual practice—I think path-breaking in their philosophical insight and noteworthy in their phenomenological closeness to the spirit of the works with which they engage—they are not as thoroughly grounded in practice, which is the great distinction of the essays in this volume. Here there is a marriage between the great philosophical themes of Hindu mythopoeisis and rite and the ground of their practice.

My discussion begins with a short consideration of certain aspects of the intellectual background to these essays, which demonstrate a holism appropriate to the mythological and ritual materials with which Handelman deals. They also manifest the development of a distinctive analytical approach in anthropology. A thesis I present is that Handelman's analytical course was virtually bound for India from the word go. That he would become fascinated with Hindu cosmology and practice was immanent in the orientation he initiated in his investigation into ethnographic materials at considerable cultural distance from that of the sub-continent. The essays here express the flowering and the potential of his thought.

A Passage to India

Handelman's journey to India and his fascination with ritual and its cosmological unfolding might be said to begin early in his career with his (Handelman 1967) encounter with the Washoe shaman, Henry Rupert. Indeed, Handelman's analysis of the Hindu materials resonates with shamanic cosmic process—the Moebius-like course into the universe of the self, its in-curving and out-curving along the multi-dimensional planes of existence, the realms of merging and shifting appearances, the emergence of the self from out of itself, and so on. The shamanic universe is a living, pulsating, embodied cosmos, and this is the hallmark of Handelman's deeply humanist perspective as it has developed. If, as I hazard, Handelman's anthropological perspective begins with the Washoe shaman, it reaches its apotheosis in and through the Hindu mytho-cosmic and ritual materials presented here. It is highly appropriate that the initial essay in this collection should concern Śiva/Murugan, the shaman archetype and generative embodiment of the living cosmos.

Almost shaman-like, Handelman's career describes a pathway through the cosmological imaginings of anthropology and the social sciences generally concerning the matter of human being. Coming to India towards the end of his shamanic journey Handelman emerges with a novel mode of anthropological understanding. To put it another way, Indian ethnography is not so much the source of his distinct mode of understanding as its realization.

My knowledge of Handelman's intellectual trajectory begins at the Department of Social Anthropology at Manchester University, where we both arrived in the mid-1960s. Max Gluckman, who had established the department, was still very much in charge, although its intellectual milieu was changing.[2] In many ways Gluckman was insisting on an orthodoxy in approach in an effort to bolster the kind of orientation—a deviation within British structural functionalism—that he felt was at the root of the innovation of his "Manchester School". The Israel research scheme that Gluckman established, which drew Don as a postgraduate to Manchester, was in part an attempt by Gluckman at reproductive confirmation.[3] I mention this because the spirit of orthodoxy that Handelman encountered was probably most intense for those employed on the Israel scheme. This was all the more so as Gluckman was, for many who encountered him, a shamanic figure in his own right. He was, some would say, "larger than life," and his anthropology embodied many aspects of his personality, into which he attempted to incorporate both colleagues and students. Handelman trained in the North American tradition of cultural anthropology and was open to the greater variation in perspective that it offered (relative to the more closed and limited scene of British social anthropology). While he had overlapping interests with Gluckman's Manchester School, he also had clear differences in orientation. In many ways, Handelman took issue with some of the key problematics of Gluckman's

---

[2] I have taken a conventional path in this presentation of Handelman's intellectual passage as exemplified in his involvement in institutional academia and in his major publications. This approach must leave out the deeply personal and the fact that Lea Shamgar-Handelman and later Galina Lindquist, sociologist and anthropologist in their own right, had a key role as intensely close intellectual companions. Lea accompanied Don on his early visit to South Asia and worked closely with him on Israel and, especially, in the development of his thought on nationalism. Lea was crucial in the formation of ideas that lead to *Models and Mirrors.* Galina became vital in more recent years and, I think, was hugely important—through her own work on shamanism, especially—in the further development of Don's ideas on ritual. The importance of both Lea and Galina for the development of Handelman's thought is evident in the record of their jointly authored publications.

[3] Gluckman, of course, built his distinctive form of analysis through his research program in Central Africa centered on the Rhodes-Livingstone Institute in Zambia (then Northern Rhodesia). The work in Zambia constituted the core of the "Manchester School" and was exemplified in the still path-breaking ethnographic work of John Barnes, Elizabeth Colson, Ian Cunnison, Bill Epstein, Max Marwick, Clyde Mitchell, Victor Turner, Jaap Van Velsen, Bill Watson and Gluckman himself. In my view, the Israel Scheme was an effort by Gluckman to further realize—and in a way repeat—the innovative frames of analysis that the Central African work largely pioneered.

deviation within British anthropology-becoming-orthodoxy and began a path of his own.

Gluckman's major methodological innovation is situational analysis (Gluckman 1958). This was aimed at overcoming the stasis of British structural functionalism connected with its emphasis on system integration and what C. Wright Mills calls abstract empiricism. This latter aspect relates to the (still persisting) tendency in much sociology to make general statements concerning social action and the structural principles relating to them without demonstrating the connection ethnographically. Often particular case material is presented as being both typical of a variety of actions that fit with pre-determined descriptive categories and as illustrative of encompassing systemic principles. Gluckman and his colleagues contested such perspectives, insisting that the ethnographic materials—the actual living practices upon which larger descriptive generalization and theory are to be grounded—should be presented and subjected to analysis, so that argument and theory could be evaluated by other anthropologists. More than this, the method of situational analysis stressed dynamics and process and a concern to enter within practice in the course of its unfolding. Accordingly it would reveal dimensions of the continually changing social whole of which it was a part, its processes of institutional reproduction and transformation. The aim was to reduce the distortion that is the fate of any analytical and theoretical orientation. Through such an approach a theoretical understanding closer to the ethnography could be progressively constructed.

Despite intentions to the contrary, situational analysis, as Gluckman and his colleagues developed it, could not escape the difficulties that it addressed. The parameters of the whole that they were examining and its theoretical comprehension were already presumed in their analysis and in their selection of events or instances of practice. This was compounded in a by-and-large externalist/objectivist orientation associated with a Durkheimian-cum-Marxist approach that Gluckman and his colleagues variously took.

When Handelman arrived at Manchester, modifications of the perspective were already underway. More phenomenologically oriented (as well as more individualist transactionalist) perspectives were being tried—in particular Goffman's symbolic interactionism, for which Gluckman himself had some sympathy.[4]

[4] Handelman informs me (personal communication) that Richard Salisbury (his supervisor for the MA at McGill University) initially impressed upon him the importance of

Handelman's PhD study of a sheltered workshop in Jerusalem (1977) is a major critique of Goffman's perspective that, in my opinion, has not achieved the attention that it warrants. Certain vital aspects of Handelman's approach were born, I think, in his critical reaction to Goffman's analytical direction; also in evidence is the innovation away from situational analysis while maintaining a connection with its spirit.

Here it must be stressed that while there was an affinity between Gluckman's Manchester situational analysis and symbolic interactionism, it was no more than an affinity. Goffman was far more interested than was Manchester in the micro-processes of social construction and repair, the dynamics and negotiation of shared social activity, the emergence and communication of meaning in interpersonal activity, and so on. While Gluckman stressed the rule of custom and institutional order, Goffman (1990, 1961) focussed on the inner process of customary or routine practices, their fragility, their negotiation to agreement, and the emergence out of them of what appeared to be ordered structures of action. In Goffman (see 2008) the ritualistic qualities of social action—their repetitive patterns—emerge from the pragmatic dynamics of social interaction. Aspects of this direction are apparent, if taken in an ultimately critical and different course, in the essays in Handelman's *Models and Mirrors* (1990).

It is worth quickly considering vital features of Handelman's approach with reference to Goffman's symbolic interactionism.

Indeed, where Goffman could be said to skate across a diversity of fragments of interactional performances from a variety of socio-cultural contexts, albeit with enormous insight, in the course of establishing his overall analytical frame, Handelman concentrates in depth on a range of specific situations and their problematics. There is a Frazerian quality to much of Goffman's work, which Gluckman's situational analysis was designed to guard against. This is a feature of Handelman's critique, although his approach does not apply situational analysis so much as develop a different approach to what he comes to define as events (see Handelman 1990, Kapferer 2010).[5] If Goffman often seems to appeal to common-sense or

---

Erving Goffman. Wesley Sharrock (now Professor of Sociology, University of Manchester) should also be mentioned. Wes was in many ways the intellectual light for a small group of postgraduate students (including myself) in the Manchester department when we arrived in the mid-sixties. Wes was then developing beyond Goffman and was encouraging the extension of our reading in phenomenology and most specifically ethnomethodology and other developments from symbolic interactionism. Handelman states (personal communication) that it was Wes who first interested him in the work of Gregory Bateson.

[5] Handelman (1990) addresses the nature of the event and different aspects of event-actualization. He takes up the issue of the event as a representation of social processes—

the stock of knowledge that his readers might already have concerning the occasions or everyday practices with which he deals, Handelman is more concerned with decentering assumptions through a close attention to the details of practice, the contexts of their emergence and, most especially, the logic of their self-constituting or organizing design and their effects.

Goffman's orientation stresses the mechanics of interactive techniques in social performance. It is largely an externalist orientation, that of the astute observer looking in from outside the situation. Handelman's orientation is far more internalist, phenomenologically close. That is, his emphasis on logics involves an entering within the processes of the formation of the phenomenon wherein the logic of its design is revealed either as a modelling of reality or else a mirroring dynamic within lived realities. In either way Handelman's approach is not founded or grounded in prior assumptions independent of the assembling dynamic or formation of the phenomenon at hand.

Moreover, Goffman's perspective risks reduction of an individualist kind that Handelman's avoids. It loses sight of the larger context or what Handelman comes to develop later as the systemic continuities into embracing processes. Goffman's perspective is organized in terms of a means-end pragmatic and strategic schema of an overarching motivational kind. With Handelman what motivations there may be emerge and change through the unfolding logic of the dynamic, integral to its design. The concept of motivation misses critical dimensions of Handelman's analytical perspective. Perhaps it would be better to think in terms of mood in a musical sense. Here I refer to the analyses in the essays in this volume where the interests, emotions, and senses of the beings in myth and the divine and human participants in rite (what I would understand as the aesthetic of their process) are vital in the design of the action. With Handelman the logic of practice has poetic energy.

---

the dominant way it has been approached in history and in anthropology—and deals with the limitation of this kind of perspective. As far as I know, no other anthropologist has attempted a careful analytical examination of the nature of the event. Certainly the main Manchester anthropologists concerned with situational analysis did not attempt this, although Clyde Mitchell (as Handelman acknowledges) was going in this direction in his seminal work, *The Kalela Dance* (1956). Evens and Handelman's (2006) edited volume on Gluckman's Manchester School, including a consideration of the history of situational analysis and some of its offshoots, presents important understandings of Handelman's own relation to the methodological development of situational analysis. Handelman's contribution concerning situational analysis and his particular orientation to micro-historical analysis is of considerable relevance to the argument in this essay and Handelman's approach in general.

I have discussed Handelman's work up to this point in relation to Gluckman and Goffman in order to highlight features of his developing approach as well as certain aspects of the intellectual context in which it was initiated.

Broadly, and in retrospect, I would say that Handelman overcame the deficiencies of the one through the other. Gluckman's emphasis on the deterministic properties of social systems was countered by Handelman's use of a Goffmanesque style of analysis that centred on the more indeterminate dimensions of social processes. Furthermore, Handelman reintroduced the systemic, very much back-grounded, into Goffman. He did so not in the more heavy-handed style of Gluckman, whereby micro-processes are conceived as the effects of macro-systemic forces and their contradictions, but through his (Handelman's) concept of the logics of design as being immanent in micro processes. The question of determinacy—how a particular logic moves into ascendancy and achieves systemic force—is for Handelman a critical issue, as his work on bureaucracy and nationalism demonstrates (Handelman 2004).

There are powerful incompatibilities in the perspectives of Gluckman and Goffman that reflect critical divisions in anthropology and the social sciences as a whole. A major aspect, which I will come to below, concerns what may be regarded as part/whole relations and an underlying dualism in perspective reflected in the oppositional discourse that directs much debate. The issue of determinacy and indeterminacy is one aspect of this discussion. Others concern the relation of the individual to society, institutional analysis vis-à-vis a focus on interpersonal dynamics, macro/micro distinctions, change versus stability, reductionism as against anti-reductionist perspectives. Manchester at the time of Handelman's presence there (indeed, the general milieu of anthropology at that time) was bound to such debates—irresolvably so.

Handelman was committed to finding a way out that, I think, is reflected in his pivotal work *Models and Mirrors*, which expresses both explicitly and implicitly the direction that is manifest in the papers published here. Many of the essays in that volume offer an alternative that is oriented away from the terms that condition the methodological incompatibilities to which I have referred and which in many ways persist. *Models and Mirrors* develops conceptualizations that render such incompatibilities irrelevant along with the dialectics of much of their oppositional discourse.

From the perspective of Handelman's current position, Gluckman and Goffman appear to be light years away, although traces continue, however transmuted. I say this to underline the significance of other major influences on Handelman. Foremost among these, in my opinion, are Victor Turner and Gregory Bateson, who can be regarded as forerunners of the analytical approach that Handelman is developing here.

From within the Manchester tradition, Turner is crucial because it is he who virtually announces the impasse that Gluckman's perspective had achieved. It was Turner who arguably developed situational analysis to its fullest extent.[6] It could go no further. Handelman's approach to events in *Models and Mirrors* represented both a way out and an approach that had immanent within it the holistic orientation that Handelman was developing largely outside the functionalism of Gluckman and others (see Evens and Handelman 2006). Nonetheless, Turner did indicate a direction towards a more unified understanding of human action—one that bridged the disciplines, especially the division between the humanities and sciences—and toward the centrality of ritual to the development of such an orientation.

Bateson, along with also offering a more unified perspective, demonstrated a route through science that broke through the hegemony of the spectre of a Durkheimian anthropological science that conceived of society as a system in itself—a system isolated from more embracing processes and other ways of conceiving. In other words, Bateson, perhaps more than Turner, opened toward other ways of understanding human action that were not necessarily centred in human being in and of itself.

Turner had left Manchester for the United States a few years before Handelman had arrived, but his presence was still palpable.[7] Turner's seminal work, *The Ritual Process* (1969), expressed his transition from the Manchester line into a new vision for anthropology. In many ways *The Ritual Process* breathes with Victor Turner's own sense of intellectual emancipation from a Mancunian orthodoxy (an attitude to which Handelman was more than sympathetic). It was a radical departure from Durkheimian and Hegelian/Marxian roots. Moreover *The Ritual Process* is a move

---

[6] It was Turner who transformed Gluckman's situational analysis into what became known as the "extended-case" method, where a series of events involving the same key participants but occurring at different moments in the historical unfolding of structural processes are examined.

[7] Handelman came into close intellectual contact with Turner some ten years later. He joined Turner and a number of other anthropologists on a kind of travelling workshop/seminar to Japan, organized by Turner and Masao Yamaguchi, in 1981. Handelman's anthropological interest in Japan and East Asia continues, perhaps as a consequence of this experience. In any case Handelman developed a close intellectual friendship with Turner, and the two gained mutual inspiration from the association.

away not only from what Turner considered to be the dead hand of British functionalism but also from the neo-Kantianism witnessed in the growing popularity of Lévi-Strauss's structuralism at the time. Turner addressed the liminal, transitional, moments of rite—the space between different moments in structural processes—as holding the key to understanding how new arrangements of meaning and structure came out of the old. In other words ritual was not the mere repetition of the same but a space for the forming of novel constellations of meaning and the creation of relations across hitherto forbidden social and cultural categories. This last was a vital dimension of Turner's concept of *communitas*, born, as he saw it, from the liminal. Handelman did not follow the Van Gennep (virtually Nietzschean) course of Turner's analysis of ritual. Indeed, he broke away from Turner's still overly linear perspective, as exemplified in his (Handelman 1990: 32–39) re-analysis of the Bemba *cisungu* initiation rite first described by Audrey Richards (1982) and influential for Turner's work, which is more strongly in Bateson's cybernetic-systems direction. Turner's emphasis on the openness of possibility—on the excess of ritual—and upon its immanent qualities—the sense that ritual is a plane from out of which multiple existential realizations are potential—is clearly of relevance to Handelman's understanding. But Bateson's (1972) orientation, I hazard, gave greater systematicity to the idea of how original possibility could be born from out of apparently repetitive processes. It also broke analytical frame. That is, it shifted course away both from the socio-centric orientation that Turner maintained and from its traces of essentialism and tension to reduction. In this regard Bateson opened an avenue whereby concepts well outside anthropology and sociology could realize dimensions of socio-symbolic processes, an avenue that was inhibited by established sociological thought.[8]

---

[8] Here I might note the distinction between Gluckman's orientation and that of Bateson. Gluckman's *Closed Systems and Open Minds* conceives anthropology, in a Durkheimian post-Enlightenment way, as a self-enclosed discipline that should interact with other disciplines on the basis of its own specialist knowledge. He championed the disciplinary territorialisation of knowledge—knowledge, that is, as a sum of autonomous parts. In other words, anthropology was a discipline that in the interests of its disciplinarity, or independent scientific re-cognition, had a particular subject matter that was not to be trespassed upon, as anthropology should not trespass in others' fields. For many in the discipline at the time, this seemed to be unduly restrictive, especially given the all-embracing disciplinary label—anthropology. Bateson argued for a discipline that was integrated into others and which bridged the humanities and the sciences. Handelman's perspective is along these lines.

Handelman (1979) addressed Bateson in relation to both play and ritual, concentrating on paradoxes of communication (as in play, whereby signs of aggression communicate their opposite and shift the register of action from non-play to play) and the complexities of feedback. He expands an understanding of Bateson's (1958) own classic ethnography of *naven* ritual behaviour through a development of Bateson's orientation. Bateson's general perspective and Handelman's engagement with it look forward to his understanding of Hindu cosmology and ritual as virtually systems of cybernetic process "that interact with themselves and produce themselves from themselves" (Kauffmann 2007).

Although Bateson is in many ways crucial to the development of Handelman's thought (this is very apparent in *Models and Mirrors*), my sense is that the cybernetic approach was possibly too mechanical. Here I suggest perspectives that had their source more in the field of quantum mechanics and in related areas of the mathematics of multi-dimensionalities, which increasingly captured Handelman's imagination. Such an orientation was already evident in *Models and Mirrors* but grew in significance through Handelman's encounter with South Asian cosmological and ritual materials. In other words, Handelman's engagement with Sanskrit and Tamil textual materials and Hindu ritual practice was a catalyst for a further development of his already independent and radical approach in anthropology. What was nascent in his earlier work—integral perhaps in his passage to India—was able to reach more complete realization.

Toward New Ontological Horizons and the Challenge of Hindu Cosmology

One way into understanding the significance of Handelman's perspective as presented in the essays here is through a consideration of the work of the cosmologist and quantum physicist, David Bohm. I well recall Don's enthusiasm upon reading Bohm (2002) and the sense that he was expressing the line of enquiry and thought that Handelman was working toward in his own way. In an important sense, India is the point where Bohm's thought and Handelman's converge. Bohm, of course, had already recognized the relevance of Hindu cosmology and philosophy to his thinking. Handelman elaborates an understanding as to why this is so; he also shows how the kind of scientifically grounded cosmological thinking that Bohm was expressing opened up avenues in Hindu thought and practice that had hitherto defied, or been occluded by, anthropological understanding.

Bohm addresses the central, famous question in quantum mechanics, one which has been of widespread intellectual fascination. This concerns the celebrated dispute between Einstein and Niels Bohr relating to determinant and indeterminate processes and the matter of underlying principles of causation. The debate, of course, had major repercussions for philosophy and reverberated through the social sciences. It influenced directions in post-modern thought and is relevant to what is being increasingly referred to as the new ontological turn in anthropology and the emergence of a new—but scientifically attested—metaphysics.[9]

Bohm resolves the Einstein/Bohr dilemma by re-imagining the cosmology/ontology in which both Einstein and Bohr's arguments hold, thus accounting for contradictions in the experimental and theoretical evidence. In Bohm's cosmic reconceptualization, both Einstein (God does not play dice) and Bohr (chance is a vital factor in cosmic processes) are right. Here I note that Bohm's cosmology is more than a paradigm shift, in the sense of Thomas Kuhn, from Newtonian physics to quantum mechanics. Rather, it is an ontological leap in conceptualization that achieves a new and radical existential unity of all that there is or might be. This is far more so than for the Batesonian cybernetic conception, which concentrates on the integrative and regulative interactive processes across systems within an overall totality. But cybernetics does not conceive of these as flowing out of a singular unity—it describes a whole that maintains the notion of functional integration of its parts, that goes well beyond the kind of functionalism of anthropology yet is still cognate with it.[10] Bohm's quantum cosmology breaks completely with such a conception. He conceives of all phenomena, what might have otherwise been understood in dualistic terms—for example, material/immaterial oppositions—as constituted in a singular multi-dimensional process wherein all forms of existential reality—from sub-atomic particles, human mental perceptions, social formations, through to the planets and stars—take form, interweave, and dissolve.

---

[9] The new ontological turn is largely epitomized in the work of Gilles Deleuze, whose approach is increasingly influential in anthropology, and, in particular, in the direction taken by Bruno Latour, Manuel DeLanda and others. Deleuze, in his critiques of Western philosophical and metaphysical traditions, has been concerned to develop a new metaphysics leading away from the past and towards science. He, too, is concerned with achieving greater unity of the humanities and the sciences.

[10] This is how Gluckman (1958) responded to Bateson, seeing in Bateson's concept of schismogenesis an association with his concepts of contradiction, conflict and moving equilibrium in relation to social change.

Bohm's cosmological/ontological reconceptualization draws a distinction between what he terms implicate and explicate "orders". That is, the fluid totality of the universe(s) of existence constitutes an implicate order from which all phenomenal existence is emergent. This implicate order becomes phenomenally explicated or actualized and more concretely materialized in a continual differentiating dynamic: the implicate order is unfolded in many explicate particular and interconnected systemic dimensions, which eventually dissolve or fold back into the continuing process of the whole from which they emerged.

Bohm's idea is of considerable relevance to Handelman and Shulman's interpretation, in *God Inside Out*, of the dice throw in the great Hindu epic, the *Mahabharata*, where the throw of the dice is both determinate and indeterminate, implicating the cosmos as it explicates (in the fall of the dice) a particular course or direction of cosmic potentiality. Handelman's essay in this volume concerning the dynamic of the Tamil cosmic hierarchy centering on Lord Murugan—its folding and unfolding and transformational shifts through multiple levels and potentials, the way the whole manifests in particular forms to dissolve back into the flow of the whole—was written well before he had read Bohm; yet it anticipates possibilities of the cosmology/ontology that Bohm articulates. Of course Bohm, like other world-changing physicists with whom he was associated (e.g. Oppenheimer who read the *Bhagavad Gita* in Sanskrit), was familiar with Hindu cosmology.

I suggest that Handelman was drawn to consider Hindu cosmology (and ritual processes) for reasons similar to those of Bohm: the dynamics that it appeared to describe meshed with the direction that Handelman was already broadly taking.

The cosmological ground of Hindu philosophy was significant for Bohm because it paralleled his argument and, I suggest, indicated the more general scientific import of his approach. While not validating Bohm's perspective in a positivist, empiricist sense, Hindu philosophy might be said to intuit the kind of cosmology that Bohm proposes as a background for theoretical understanding—indirectly rather than directly. Let me explain, because here I think there is a link between Bohm and Handelman.

The Hindu concept of *maya* or illusion has clear bearing on Bohm's distinction between the explicate and the implicate—the former being an irreducible manifestation of the continuous changing flow of the cosmic whole. Furthermore, as Bohm might argue, *maya* is a perceptual illusion that springs from the reality of existence but obscures or shrouds its more profound nature. Very possibly, Hindu philosophy based in the *Veda* gave

Bohm a way to re-imagine the cosmology to which his scientific evidence pointed. This was so because its difference, like all difference, in Bohm's holistic orientation (indeed, in all radical holism) indicates a more profound unity. In the differences, in the contradictions, in the disjunctions, lie clues to what this unity might be. But given the vast implications of Bohm's project—the understanding of a physics whose dynamics is at the heart of existence in its entirety, including the existential realities of human beings *even as they imagine these perceptually and conceptually*—all formations within existence contain dimensions of the implicate order of the cosmos. That is, Indian philosophical or mythopoetic reflections on reality are not just good to think with in a Lévi-Straussian sense, a way "to think outside the box" or to break out of the hermeneutic circle that is integral to all holism; they themselves manifest the very nature of the existential reality they reflect.

In this context, I note, Bohm's physics and the cosmological re-imagining that it indicates have been labelled "mysticism" of a neo-Platonic kind. But this is a misunderstanding, regardless of whether Bohm's cosmology is accurate or not (for as Bohm would maintain, the nature of the universe is an enduringly open question that conditions any scientific attitude and its continually changing understanding and reconfigurations). Unlike neo-Platonism, Bohm's perspective is not grounded in religion and involves no *deus ex machina.* If it is "mystical," it rather bears some connection within esotericism (in the *Veda,* or *Kabbalah,* or even more so Spinoza's philosophy), which argues for deeper "truth" as secreted (hidden and released) within the formations and reflections that claim truth but also conceal it (that indeed must conceal it). It is in the *concealment of unconcealment* that another far more accurate or approximate truth resides. Bohm's differs from this mystical/esoteric tradition in that his philosophy is dependent on science (on the testing of evidence in relation to cosmological contingency and potential). That is, Bohm breaks with metaphysical traditions whose conceptual principles or parameters of understanding remain unchanging.

My attempt to connect Handelman's thought to Bohm's does not in any way intend to confine Handelman within Bohm's cosmological/ontological terms. Nonetheless, I think the brief discussion of Bohm does bring out key aspects of Handelman's direction that finds some of its inspiration from science, and especially theoretical physics. This is so, I hasten to add, not in the sense of seeking an authorization in science. As with the best anthropologists, the authorization of the approach is in the ethnographic evidence itself. For Handelman, science is good to think with neither more nor less than myth, for example, is for Lévi-Strauss. With Handelman, myth and ritual, on the one hand, and science, on the other hand, participate together in the conceptualizations they offer toward comprehending human existential processes and the way human beings approach their realities. Scientific concepts such as those of Bohm (and others, especially those relating to post-Newtonian physics: see e.g. Barad 2007) enable Handelman to think *through* the masks of Hindu cosmological terms and processes—through the resistance that their direct appearance to anthropological perception may present—to reveal a more phenomenologically attuned understanding of the kind of logic or design that they are expressing. Simultaneously, I think, Hindu practices reveal the possibilities of the methodological perspective that Handelman has been unfolding throughout his career for an understanding of human practices in general.

Handelman and a New Holism for Anthropology

Handelman's orientation achieves significance in comparison with that of Louis Dumont, who is an absent presence in these essays. Dumont's shadow is apparent in "Myths of Murugan." This essay can be read as an indirect critique of Dumont or perhaps, more accurately, as opening up a space of analytical interpretation of Hindu ethnography between Dumont and his critics. While Dumont is chiefly concerned with the social processes of caste and kinship, his argument hinges on a particular presentation of the Hindu—principally, the Vedic-hierarchical cosmological scheme of things. Religion and ritual define the holistic principles of social order; Dumont begins his famous work, *Homo Hierarchicus* (1980), with an analysis of the logic of this scheme. He argues that the political or power (in the shape of kingship or in the *varna* category of *kshatriya*) is subordinate to the religious or ritual (in the shape of the priest or *brahman*). This argument has stimulated widespread debate; the essays here can be regarded as an intervention within it. This feature is an important aspect of the two essays on ritual, especially the Paiditalli essay on the growing of the state. They explore certain features of the relation between ritual practice and the social and political but arrive at a different understanding of cosmic, social and political integuments than that of Dumont.

These essays expand dimensions of Hindu cosmological realities that are important to Dumont's argument (and to other positions critical of Dumont). However, they take a radically different analytical course, demonstrating the distinct holism that Handelman has been developing. To some degree they could be seen as resolving disagreements in interpretation between Dumont and his critics and incorporating them in a different cosmological vision. McKim Marriott (1976), a major critic, stresses Dumont's inattention to fundamental features of Hindu philosophy and cultural practice with which Handelman engages as integral to his own analysis. For Dumont the kind of approach that McKim Marriott epitomizes constitutes an anti-holist reductionism that is compounded by an intense relativism. Thus Marriott exemplifies a holism that ultimately founders on a part/whole dilemma whereby the whole is a formation arising out of the interweaving or summation of its parts. As I have said, Handelman develops a holism, that redresses the problem that Dumont discovers in Marriott as well as difficulties that remain in Dumont's holism. The whole is neither generated from the interrelation between its parts nor is it the sum of (or greater than the sum of) its parts but, rather, is prior to that which is created or generated from it. Handelman's holism is outside the dualism that continues to bedevil Dumont's despite his attempt to transcend it.

Although Dumont was eager to criticize the dualism and ethnocentricism that marked much research on India, he often seems to have exacerbated it. In some respects, this is because his perspective both embeds the very dualism that he is concerned to transcend and perpetuates some of the dualist oppositions that are intrinsic to the sociology out of which his anthropology developed. Moreover, Dumont's approach is vulnerable to charges of essentialism and evolutionism or primordialism, even though these are far from his intention. They, too, are a legacy of the sociological tradition in which Dumont is located.

Handelman's conceptual development of more (scientifically) neutral terms, such as asymmetry and, in more recent work, fold (and fractal), avoids the conceptual ambiguities of Dumont's chief concepts of hierarchy and encompassment. The very ambiguity of these concepts exposes Dumont to criticisms of the cultural-centric kind that he critically addresses and attempts to avoid.

Dumont's concept of hierarchy is the dynamic and overarching, unifying (cosmological) principle of his analysis based in the opposition between the pure and the impure. He asserts the primacy of this opposition specifically in relation to understanding caste. This is hotly debated by other Indianists (e.g. see Raheja 1988). It is this position which exposes him to the criticism of essentialism.

The holism and the dynamic of Dumont's argument rests in his conception that the pure and the impure are defined in relation to the whole (as right is to left in relation to the human body) and thus inseparably entwined—the pure being valued over the impure, so that they describe a relation of hierarchy. It is their hierarchical integrity together with their vital incompatibility (the one excludes the other) that comprise the dynamic force of differentiation of caste in the continuing formation of social relations, together with their political and economic effects. I suggest that this conception contains what amounts to an aporia of dualism in Dumont, compounded in the dyadic nature of his analysis.

The argument that Dumont develops also harbours other dualisms founded in the oppositional quality of his analysis and, paradoxically, an ideological commitment vital within the very dualism of Euro-American thought. I highlight this point because Dumont (see Dumont 1986, 1994; also Kapferer 2013) is concerned to overturn the authority of Euro-American social science, since it overvalues—or is over-determined by—the history and philosophy from which it emerges. Dumont is strongly anti-reductionist, and his structuralism involves an antagonism to subjectivist and frequently individualist (or atomic) kinds of anthropological understandings, including those of phenomenology. But such an oppositional position is, of course, part of the philosophical history of which he is a part and is an ideological assertion within its terms. Far from transcending dualism, Dumont promotes it and, indeed, guarantees the ideologically driven criticisms that he receives.

All that I have said serves to mark Handelman's distinction. In a sense, the holism that Handelman advocates encompasses Dumont and avoids the difficulties that Dumont has encountered while realizing some of his objectives.

As I have already suggested, Handelman's orientation does not rest on relatively stable dualities or oppositions of a fundamental nature. The dualisms, such as they may be, come into phenomenal play through the dynamic of the whole. They are not its essential premise. There are multiple planes of opposition in Handelman's analysis, their terms changing in quality and in register. (The critical religio-philosophical concept of the *gunas*, which Dumont ignores, are thoroughly incorporated into Handelman's analysis.)

Handelman stresses the vital dimension of the dynamic of the whole, involving shifts in the character or qualities of whatever comes into and out of opposition in the process of a cosmic unfolding in myth and rite. The oppositions are not stable, and it is their change which is critical to his understanding of the cosmology implied by the practices that Handelman addresses.

Where Dumont's dualism stresses a singular axis or plane of opposition, Handelman's is more topographical; the coordinates are continually changing, as is evident both in his analysis of the Murugan myths and in his discussion of rite. Altogether, Handelman's approach is multidimensional, virtually holographic (a notion that Handelman engages) in contrast to the two-dimensional, flat perspective of Dumont. What may be described as Handelman's quantum anthropology is thoroughly appropriate to the aesthetic and sensual depth of realities as they are presented in the poetics of text and as they unfold in the drama of rite. There is an identity between Handelman's analytic method and the phenomena that he addresses.

The hallmark of the analysis in these essays is what could be called their phenomenological nearness. The analysis involves entering into the phenomenon—exploring it from the inside out, as it were—rather than from an outside, objective distance as in much structuralist anthropology and certainly in that of Dumont.[11]

The anthropologists Victor Turner and Clifford Geertz argued against structuralist orientations largely because the latter involved a destruction of the phenomena to be grasped and, therefore, the loss of critical dimensions of their understanding. They stressed the aesthetics, the sensual subjectivities of mythic and ritual processes. This, I note, is not absent from the structuralist anthropology they mostly opposed—even if it is from Dumont. Lévi-Strauss attempts methodologically to align an aesthetic and subjectivist entry into the phenomena he analyses with a more objectivist understanding. This is particularly evident both in his major corpus on myth and, significantly, in his analysis of shamanic ritual.

Lévi-Strauss's discussion of the logics of myth equates their design and movement with that of music—where, in his view, art and science discover their essential unity (see Lévi-Strauss's 1990: 660–7 extended discussion of Ravel). For him music bridges difference and becomes the

---

[11] The observation that Handelman's approach is not flat, more "in the round" relative to that of Dumont, needs much greater explication. I wish to draw attention to his concern to enter into the realities concerned through the perspectives of the different participants, whether they be gods or human beings. These beings are presented by Handelman (also in the two jointly authored books by Handelman and Shulman) in terms of their changing perspectives (and their embodied constitution) on the emerging situations of which they are a part. The orientation is very much in line with that which has been labelled "perspectivism" in anthropology best exemplified currently by Viveiros de Castro (2012).

methodological and affective means for entering into otherwise alien cosmological and mythological realities. Lévi-Strauss's discussions of shamanic birthing and healing virtually bring together his structuralism with Merleau-Ponty's attention to the phenomenology of the body. However, overall the Lévi-Straussian orientation remains schizophrenic, unable to transcend the subject/object dualism upon which it is based, oriented as it is in the direction of Kant's failed resolution in the *Critique of Judgement.* Ultimately, with Lévi-Strauss the objective subdues the subjective or pushes the subjective away. The same, but in reverse, can be said of Turner and Geertz who, despite orientations away from Kant and their strong humanist and phenomenological emphasis, sustain and even promote the Kantian dualism to which they are opposed: Humanism is opposed to Science.

In Handelman Science and Humanism are effectively brought together (the sensual is not suppressed) in what amounts to a new phenomenology, which has its roots in scientific evidence rather than in the metaphysics of religion. This is the direction of Bohm and, incidentally, others such as Steven Rosen (1994), who develops the concept of the Moebius principle to reconceive apparent asymmetries and contraries as united in symmetrical processes. Rosen (2008) sees in Merleau-Ponty's phenomenology a true transcendence of Kant and an existential unity with conceptions founded in a science-informed cosmological holism. I mention this both because the Moebius concept is explicitly vital in Handelman's perspective and to underline key aspects of Handelman's phenomenological reorientation, which is concerned with the dynamics of perception and the senses as integral—indeed inherent within—the logics of the whole.

Thus where Dumont's holism develops in terms of a series of complementary oppositions, Handelman's Moebius holism stresses the interweaving of that which may otherwise be definitively opposed. This gives a different slant to Dumont's perspective. For example, Dumont argues that in Hindu India religion opposes and encompasses the political—the religious is the condition of power. Much debate has followed on this assertion. Handelman et al in the essay on the growing of the state do not follow an oppositional discourse but show the complexity of the interfusion of the religious with power—or, more accurately, how power is not the opposite of the religious but rather emerges in a singular cosmic dynamic. Power is simply a critical dimension of unity. Similarly, Handelman's discussion of religious festival does not oppose the religious or ritual to the social (or view one as the image of the other in a Durkheimian sense) but conceives them as intermeshed and different or relative possibilities of a single process. Handelman, in these regards, might be said to extend Dumont's argument—he reveals the particular integration of socio-political processes within the religious and cosmological whole, but from a different angle. His analysis may also add to the understanding of the resilience of the overtly religious in social and political processes in India.

Dumont argued that socio-cultural practices in India refracted a non-dualist holism apparent in the sacred texts. He developed a conceptual methodology that he claimed was appropriate to such a holism, which furthermore could inform the development of a rigorous scientific comparative anthropology. Dumont asserted the radical thesis that the holism that he advocated overcame—as it revealed—the ethno-centricity of Western social scientific conceptualizations. Via India a truer understanding of humanity in general could be attained—rather than via the dualistic philosophical and analytical systems that were the heritage of Western historical systems and the proclaimed modernity of its post-Enlightenment thought. Dumont's perspective is past- and traditionalist-oriented. Indeed, Dumont's (1986, 1994) approach to India and to the world in general could be described as entropic; modernity and the present are seen as forms of devolution from an original holist universe. Moreover, paradoxically, Dumont can be seen as reversing the terms of social science discourse. India as the object of Western social scientific understanding—a plane for the demonstration of the potency of its concepts—becomes the source of new conceptual construction potentially more powerful than those hitherto applied to it.

Handelman's approach also recognizes in Hindu cosmology and practice a holistic dynamics, which corresponds—in its multi-dimensionality, in the multiverses of its complexity, in the sensuality of the logics of its processes, and in the integral unity it describes—with the new cosmologies of science that have broken free from ancient cosmologies.[12] In this sense Hindu thought and practice are as much future-oriented as traditionalist. Most importantly, there is in both the identity and the difference between Hindu thought and practice, on the one hand, and those of quantum science, on the other hand, a way in—revealing the existential nature of Hindu realities and their imaginal and experiential force as well as the

[12] Gilles Deleuze in *Cinema 1* (1986) states clearly that the scientific/technological age—in which his philosophical re-orientation is located—marks a break with the relatively fixed and unchanging cosmologies upon which preceding metaphysics was largely based.

potential of current scientific thought for an anthropological understanding of other domains of human existence.

Unlike Dumont, Handelman does not make India the paradigm and the source of anthropological understanding. He eschews the dialectics of so much anthropological discourse based in the concept of the Other as the ground for a reconceptualization of the social scientific Self and the comparative investigation of the nature of human realities. What he does do—and here he is in agreement with Dumont—is to assert that human realities must be grasped through their own processes and systems of self-conceptualization and construction. This commitment accounts for Handelman's interest in ritual. It is an anthropological commitment that may explain the enduring interest ritual has for anthropologists generally. However, Handelman's scientific holism does not grant any particular socio-cultural reality or practice any greater authenticity than any other, as I believe is the implication in Dumont (and in much anthropology). Dumont is strongly anti-relativist but unfortunately provokes relativism in his method and thus encourages thinking in terms of polarities. This is not the case with Handelman, who has developed a conceptual orientation that is equally applicable across a diversity of human practices, revealing the distinctiveness of their particularity within a holism that can truly make universal claims.

Conclusion

I have discussed the essays here in terms of the arc of Handelman's anthropological intellectual journey. He arrives at his analyses of Indian materials through critical investigation of ethnography and various conceptual approaches in contexts well outside India. But it is as if the understanding that he was developing virtually ordained his analytic encounter with the ethnography that is presented here. The work here is a kind of demonstration of his orientation-in-the-making.

Handelman's method is a work in progress. I have referred to some of the important lines of its relatively early development and emergence through a critical analytical evaluation of ethnographic materials—both his and others'. In my view Handelman, more than any other anthropologist, has been concerned with the re-analysis of very diverse ethnographic materials and, as in the case of the essays here, with revisiting the contexts of others' analyses. In doing so he has usually arrived at altogether new interpretations of the evidence and a re-configuration of analytical or theoretical understanding.

From the start of his career, Handelman was never prepared to accept conventional or established understanding or newly fashionable views, no matter how apparently radical, without careful examination of the evidence. Handelman is not doctrinaire, like those who might opt for a Marxist program of analysis at one time or for Lévi-Straussian structuralism at another. The problem with such orientations is that they tend to shut out possibilities rather than being continually open to potential, hence to the constant movement from one fashion to another in anthropology. Their very design, Handelman might say, defines the terms of their limitation. That is, the finitude of their organizing principles and procedures exacerbates a tension toward validating their legitimacy in and of themselves over against their capacity to discover and reveal. They become phenomena for confirmation "in their own right" and lose their ability to enter into phenomena as tools in an attempt to expose the particular and perhaps also the general nature of the processes that these diverse phenomena express. The great distinction of Handelman's methodological perspective for anthropological understanding, indeed eventually for a comparative comprehension of human diversity, is that his analytical "system" is in a state of what could be called *perpetual immanence.* Thus Handelman's approach is in a continual process of adjustment and building as he encounters new phenomena. These essays, providing as they do original insight into cultural practices that have long fascinated a wide spectrum of scholars, are stages in Handelman's continuing expanding intellectual horizon.

Bibliography

Barad, Karen, 2007. *Meeting the Universe Halfway: Quantum physics and the entanglement of matter and meaning.* Durham, Duke University Press.

Bateson, Gregory, 1958. *Naven* (2nd edition). Stanford, Stanford University Press.

——, 1972. *Steps to an Ecology of Mind.* New York, Ballantine.

Bohm, David, 2002. *Wholeness and the Implicate Order.* London, Routledge.

Castro, de Viveiros, 2012. *Cosmologies: Perspectivism. Hau* (open journal) vol. 20.

Deleuze, Gilles, 1986. *Cinema 1: The Movement-Image.* Translated by Hugh Tomlinson and Barbara Habberjam. London: Continuum.

Dumont, Louis, 1980. *Homo Hierarchicus.* Chicago, Chicago University Press.

——, 1986. *Essays on Individualism.* Chicago, Chicago University Press.

——, 1994. *The German Ideology.* Chicago, University of Chicago Press.

Evens, T. M. S. and Don Handelman eds., 2006. *The Manchester School: Practice and Ethnographic Praxis in Anthropology.* New York, Berghahn.

Gluckman, Max, 1958. *The Analysis of a Social Situation in Modern Zululand.* Manchester, Manchester University Press.

——, 1967. *Open Minds and Closed Systems: The Limits of Naivety.* in *Social Anthropology.*

Goffman, Erving, 1990. *The Presentation of Self in Everyday Life*. London, Penguin.
——, 1961. *Encounters*. London. Macmillan.
——, 2008. *Interaction Ritual.* London, Aldine Transaction.
Handelman, Don, 1967. "The Development of a Washo Shaman." *Ethnology*, 6: 444–464.
——, 1977. *Work and Play Among the Aged: Interaction, Replication and Emergence in a Jerusalem Setting*. Assen: Van Gorcum.
——, 1979. "Is Naven Ludic?—Paradox and the Communication of identity." *Social Analysis*, 1: 177–191.
——, 1981. "The Ritual Clown: Attributes and Affinities" *Anthropos*, 76: 321–70.
——, 1983. "The Madonna and the Mare: Symbolic Organization in the Palio of Siena." in *Spectacle—An Anthropological Inquiry*, edited by M. Yamaguchi, Tokyo: Sanseido, pp. 153–184 (in Japanese).
——, 1990 (1998). *Models and Mirrors: Towards an Anthropology of Public Events*. Cambridge: Cambridge UP (2nd ed.) Oxford, Berghahn.
——, 1993. "Passages to Play: Paradox and Process." *Play and Culture* 5: 1–19.
——, 2004. *Nationalism and the Israeli State*. London, Berg.
Handelman, Don and David Shulman, 1997. *God Inside Out: Śiva's Game of Dice*. New York, Oxford University Press.
——, *Śiva in the Forest of Pines: An Essay on Sorcery and Self-Knowledge*. New York, Oxford University Press.
Kapferer, Bruce, 2010. "In the Event—Towards an Anthropology of Generic Moments." *Social Analysis* 54: 1–27.
——, 2013. "How Anthropologists Think: Configurations of the Exotic" (Huxley Memorial Lecture, December 2011). Journal of the Royal Anthropological Institute, in Press.
Kauffman, Lou, 2007. Cybernetics Discussion Group. Wikipedia System Science. CYBCOM Archives <CYBERCOM@HERMES.GWU.EDU> 21 September. Accessed 15 July 2013.
Lévi-Strauss, Claude, 1981. *The Naked Man* (*Mythologiques* Vol. 4). Chicago, University of Chicago Press.
Marriott, McKim, 1976. "Hindu Transactions: Diversity without Dualism" in *Transaction and Meaning: Directions in the Anthropology of Exchange and Symbolic Behavior*. Edited by Bruce Kapferer. Philadelphia, Institute for the Study of Human Issues, 109–142.
Raheja, Gloria, 1988. *The Poison in the Gift*. Chicago, Chicago University Press.
Richards, Audrey, 1982. *Chisungu: A Girl's Initiation Ceremony among the Bemba of Zambia*. London, Tavistock.
Rosen, Steven M., 1994. *Science, Paradox, and the Moebius Principle*. Albany, State University of New York Press.
——, 2008. *The Self-Evolving Cosmos: A Phenomenological Approach to Nature's Unity-in-Diversity*. Singapore, World Scientific Publishing.
Turner, Victor W., 1969. *The Ritual Process*. London, Routledge.

CHAPTER TWO

MYTHS OF MURUGAN: ASYMMETRY AND HIERARCHY IN A SOUTH INDIAN PURANIC COSMOLOGY*

Don Handelman

Myths of creation tend to be especially informative about the organization of cosmology. Such accounts of emergence often bare cosmological principles that, in other kinds of myth, are condensed or summarized. In this regard, Tamil Shaivite conceptions of the divine seem contrary. On the one hand, there is god, the distant and sublime ruler, who is identified with the cosmos and responsible for its regulation and maintenance. This sense of the divine is central to the theology of Śaiva Siddhānta.[1] On the other hand, there is the wholly compassionate divinity who is close to, and is intimately involved with, his devotee. Remote ruler or ardent lover, transcendent or immanent—these abstractions appear mutually incompatible. However, the interpretation offered here of portions of a Tamil text, the *Kandapurāṇam (Kantapurāṇam)*, suggests that these aspects are located along a cosmic continuum that depends for its coherence, in part, on relations of hierarchy and encompassment.[2]

The *Kandapurāṇam*, likely composed in the seventeenth century, is a classical rendition of the Murugan myth, one prevalent in the northern Tamil country.[3] It tells the story of the god, Skanda, but bears no relation

---

* This work is in memory of Victor Turner—friend, comrade, and bhakta of the human spirit. I am grateful to Bruce Kapferer who enticed me to Murugan's thicket, to David Shulman who made me welcome and warm within, and to A. K. Ramanujan who exposed yet more deeply rooted undergrowth. Errors of direction are my own. Previous versions of this essay were read to the South Asian Studies Colloquium of the University of Wisconsin—Madison in May 1983 and to the Third Axial-Age Conference, Bad Homburg, July 1985. My thanks to the participants for their comments.

[1] Proper names and other terms are presented in phonetic approximation to spoken forms. The first occurrence of each Tamil term is followed by an exact transliteration and, in turn, by a transliteration of the Sanskrit equivalent, where relevant. The first occurrence of Sanskrit terms, especially those of less common usage, is transliterated.

[2] These terms are used in the general sense of those of Louis Dumont. See his *Homo Hierarchicus: The Caste System and Its Implications* (London: Weidenfeld & Nicholson, 1970), and his "The Anthropological Community and Ideology," *Social Science Information* 18 (1979): 785–817, esp. 809.

[3] The *Kandapurāṇam* was composed by Kacciyappacivacariyar in the Kumara (i.e., Murugan) temple in Kancipuram. For the substance of this text, I rely on the extensive

1. Murugan, south Indian bronze, c. 16th century.

to the better-known Sanskrit *mahāpurāṇa*, the *Skandapurāṇa*.[4] Skanda is known in South India by a host of names and is said to be perhaps the most popular of deities there today.[5] Here, to avoid confusion with the Sanskritic Skanda, the deity will be called Murugan (Murukaṉ) to denote his more transcendent configuration, and Shanmugan (Caṇmukaṉ/

---

summary by R. Dessigane and P. Z. Pattabiramin, *La légende de Skanda: Selon le Kandapurāṇam Tamoul et l'iconographie* (Pondicherry: Institut Français d'Indologie, 1967). Other summaries of portions of this text are given in David D. Shulman, *Tamil Temple Myths: Sacrifice and Divine Marriage in the South Indian Śaiva Tradition* (Princeton, N.J.: Princeton University Press, 1980), pp. 275 ff.

[4] Shulman, pp. 30–31.

[5] Fred W. Clothey, *The Many Faces of Murukaṉ: The History and Meaning of a South Indian God* (The Hague: Mouton, 1978).

Āṟumukaṉ; Sanskrit Ṣaṇmukha, the Six-headed) to denote his more immanent condition.

The story of the *Kandapurāṇam* is one of cosmogenesis. It begins with a form of cosmic destruction and then relates the reconstitution of cosmic order. The story is also one of the ontogenesis of Godhead, of the creation and maturation of Shanmugan ~ Murugan, the restorer of cosmic order and its divine ruler.[6] Thus this myth is one of the taking apart and the putting together of the phenomenal world. Since this story is lengthy and complicated, a brief synopsis of the plot is provided here to alert the reader to its progression.

The divine pair, Shiva and his consort, separated. He granted to the enemies of the gods, the asura brothers, the powers to rule the cosmos in their own destructive way. Oppressed, the gods pleaded with Shiva to have a son who would rescue them. Shiva created Shanmugan ~ Murugan, the deity, to destroy the asuras. Shanmugan was a playful, capricious child of great force. Shiva's consort, Parvati, rejoined him and tended to the child. As the young deity grew into manhood, Shiva armed him with weapons that included the terrible lance, the *vel*, which could destroy all illusion. Shanmugan annihilated the asuras and was worshiped as the supreme deity.

In this text, the asuras are constituted of qualities that are the obverse of those of the deity. Their mother, Māyā, was a mistress of illusion. To birth them, she seduced the sage Kashyapa (Kācipaṉ; Sanskrit Kāśyapa). The asuras were taught the truth of illusion and the materialist values of wealth and power. To accumulate these, they became great ascetics and loving devotees of Shiva. He responded to their self-sacrifice with boons of terrible powers, which they used to conquer the gods and to overturn the proper ordering of the cosmos, until their defeat by Shanmugan.

After slaying the asuras, the deity married Devayanai (Tĕyvayāṉai; Sanskrit Devasenā), the daughter of Indra, the king of the gods, and the couple lived regally in the abode of the gods. In the meantime, in the hill country, the glance of a sage, Shivamuni, impregnated a doe who gave birth to a daughter. She was adopted by simple hunters and given the name Vaḷḷi. Hearing of the girl's beauty, the deity journeyed to see her. Smitten, he courted and won her love. They married and returned to Devayanai in the abode of the gods, where the trio lived thereafter in harmony.

---

6 The sign " ~ " between terms (e.g., Shanmugan ~ Murugan) indicates their relationship to be ongoing, continuous, and processual.

Wendy O'Flaherty makes the point that Indian texts tend to be self-referential. Indeed, reflexivity, considered here as the consciousness of self-consciousness, thrives on contrast, contradiction, and opposition. The acute asymmetries that emerge through its story make the *Kandapurāṇam* a highly reflexive text. But, beyond this prevalent attribute, the very genesis and maturation of the divine in this cosmology depend on the attainment of reflexive capacities that are contrasted with those of the merely reflective, considered here as self-consciousness. I will argue that the evolution of divinity in this myth is the movement from the lower-order reflective to the higher-order reflexive, a metamorphosis to self-realization that is integral to a deity who encompasses the cosmos in his being.

Yet, in this system, god also devolves toward his devotee, and this entails a loss of reflexive capacities. The fully reflexive being is transcendent, regally and cooly distant from his worshipers. The immanent and interventionist deity, who responds to the devotion and love of his devotees, is dangerously reflective. But neither at the zenith of the evolution of divinity nor at the nadir of its devolution is there a complete solution to the relationship between god and man, between the divine and the soul. And each extreme leads, hypothetically, to the virtual dissolution of the divine and therefore to the destruction of the ordered and functioning cosmos. Therefore, the cosmic order that emerges through this text is one that moves, ascending and descending through the continuum of hierarchy without, as it were, drifting into its dangerous extremities where the deity loses his reflexive capacities. This cosmic organization may be summated as one of ongoing self-transformation (which includes also the transformation of the self).

The first section of this essay discusses the fragmentation of the cosmos, in which Shiva, the embodiment of the male principle, is separated from Shakti, the female principle. The asymmetry of Shiva produces Shanmugan ~ Murugan, the deity who eventually restores the cosmic balance of forces by bringing Shakti back under his control.

The second section takes up the asymmetry of Shakti, apart from Shiva, which produces the asuras and their vision of a materialist cosmos that wrecks the processes of karma, of death and rebirth, and the order of the dharma. With the triumph of the deity, this version of the world is destroyed and its forces are assimilated to those of the deity. In turn, this completes the evolution of the deity as the divine encompassment who forms the upper reaches of the cosmic continuum.

The third section relates the devolution of the deity, from encompassment to immanence, through his marriage to two consorts, or shaktis.

With these marriages he completes the remainder of the hierarchical continuum, which he then descends and ascends through the influence and mediation of his consorts. He now is constituted to draw near to his devotees and to regally distance himself from them.

The fourth section introduces some factors that continuously unbalance the triad of deity and consorts. These factors help to keep him, and the cosmos identified with him, in an ongoing condition of self-transformation. The last section, the conclusion, discusses the kind of hierarchy implied by the text and the significance in this hierarchy of the structures of monad, dyad, and triad.

From Fragmentation to Encompassment: The Asymmetry of Shiva

Let me begin with an episode that occurs toward the end of the text (in canto 5) but that serves as a prologue to the unleashing of asymmetrical forces in the relationship of Shiva (the embodiment of the male principle) and Shakti (the female principle). Uma (the Goddess, the embodiment of the female principle) was incarnated as the daughter of Daksha (Tamil Takkaṉ), and Shiva agreed to wed her. During the nuptials, Shiva disappeared and was cursed by Daksha. Later on, Shiva took Uma away as his bride. Daksha held a great sacrifice to which he invited all the gods, with the exception of Shiva. Shiva destroyed the sacrifice, beheaded Daksha, and threw his head into the sacrificial flames.

Near the beginning of the text, Uma told Shiva that she hated her body, for so long as she remained within it she was the daughter of Daksha. He told her to be reincarnated as the daughter of Himalaya. She left Shiva and became the child Parvati, who performed austerities (*tapas*) to attract Shiva as her groom. Separated from Uma, his shakti, Shiva withdrew himself into austerities and meditation. In order to exact vengeance on those who had participated in Daksha's insult to him, he granted the asuras the power to conquer the gods.

The Death of Kama: Asceticism Ascendant and the Nadir of Shakti

Suffering under the rule of the asuras, the gods forced Kama, the embodiment of desire, to entice the meditating Shiva from his austerities, so that he would marry Parvati and have a son who would destroy the antigods. With shivering trepidation, Kama shot his arrows of lust into Shiva. Shiva opened his third eye that, as O'Flaherty notes, is an "ascetic and anti-erotic force,"[7] and burned Kama to ash. Shiva returned to his meditation. All the gods assembled before Shiva's palace and praised him. He then became warmer toward them, left his austerities, married Parvati, and agreed to have a son. He restored Kama to life, but as an invisible being whom only the latter's wife could see. Yet no son was born, and a deputation of gods came once more to Shiva. They found Parvati and him sitting in stillness on separate thrones. Again they entreated him to have a son.

The great cycle of asymmetry in this myth, of the taking apart and the putting together of cosmic order, begins with the separation of cosmic forces signified respectively by Shiva and Uma. Together they are fertile, generative, and equilibrating, but apart they are potentially destructive.[8] Divided from one another, Shiva and Uma/Parvati accumulate the ascetic "heat" of austerities. Shiva, apart from his shakti, fails to maintain the order of the dharma, which is wrecked by the asuras. The dominance of imbalance is heightened with the burning of Kama, for he signifies the powers of erotic desire, and probably that of fertility, that keep the karmic cycle of death and rebirth moving in the world of saṃsāra.[9] The power of asceticism is ascendant. Although Shiva relents and resurrects Kama, the latter returns to only an invisible shadow of his former self and strength. And although Shiva weds Parvati, their joining apparently is not consummated and so is infertile.

In terms of the structure of the story, the death of Kama—the destruction of erotic desire—is crucial. For this fracture between fertility and asceticism, between karma and stasis, continues until the death of Surapadman (Cūrapaṉmaṉ; Sanskrit Śūrapadma), the last and the most powerful of the asura brothers who rule the cosmos. In contrast to the power of Kama, that of the asuras signifies the aggrandizement of materialist desire, of lust without fertility, of Shakti without Shiva. On the side of the gods, the ineffectuality of Kama signifies the weakness of Shakti in relation to Shiva. For, as his ascetic power grows, Shiva destroys, as it were, his own erotic power of normal fertility and therefore that, too, of the

---

7 Wendy Doniger O'Flaherty, *Asceticism and Eroticism in the Mythology of Śiva* (London: Oxford University Press, 1973).

8 See C. J. Fuller, "The Divine Couple's Relationship in a South Indian Temple: Mīnākṣī and Sundareśvara at Madurai," *History of Religions* 19 (1980): 321–48.

9 See Frederique A. Marglin, "Types of Sexual Union and Their Implicit Meanings," in *The Divine Consort: Rādhā and the Goddesses of India*, ed. J. S. Hawley and D. M. Wulff (Berkeley, Calif.: Religious Studies Series, 1982), pp. 305–7; and Robert Perinbanayagam, *The Karmic Theater: Self, Society, and Astrology in Jaffna* (Amherst: University of Massachusetts Press, 1982), p. 68.

gods. As will become clear in the following section, complementing this development on the side of the asura brothers, their mother, Māyā, develops to its extreme the power of shakti uncontrolled by Shiva. In a manner typical of the consequences of asymmetry in this myth, the negation of erotic desire on the side of Shiva and the gods produces its unbridled but infertile magnification on the side of the asuras. The effect is the destruction of the karmic cycle and the tearing apart of dharmic order.

The Creation: A Chain Reaction of Fragmentation

The birth of the deity, Shanmugan ~ Murugan, is redolent with asymmetrical power, bursting energy, and thwarted fury. Shiva, wrapped in austerities, responded to the petitions of the gods. His aloof, withdrawn, and singular countenance suddenly sprouted six heads.[10] From the third eye of each head there shot forth a fiery spark, the seed of Shiva. The six sparks spread everywhere with terrible heat. The air dried, the seas lost their waters, and the dessicated earth split into deep crevasses. The sparks were carried in agony by Vayu (wind), Agni (fire), and the Ganges (water), to Lake Śaravaṇa and were deposited there. All ran in terror from their burning heat. Parvati's anklet contained nine precious stones. As she fled, the anklet broke, spilling the gems.

Within the lake the six sparks momentarily took the form of a child with six heads and twelve arms. Then this form burst asunder into six boys, each of whom was nurtured by one of the six Pleiades maidens (*Kṛttikās*). Parvati, denied her natural motherhood, cursed the wives of the gods to be barren. Each gem of Parvati's anklet reflected her visage. Shiva commanded these reflections to come forth. From the nine stones there emerged nine shaktis. As his gaze fell on them they became with child. The enraged Parvati cursed these shaktis to an unnatural period of pregnancy. Later, unable to bear their condition, the shaktis begged Shiva's intercession. He commanded Parvati to lift her curse, and soon each shakti bore a fully grown son. These offspring became the generals

---

10 Shiva often is depicted with five heads. Here he prefigures the form of his son. This fits well with his pronouncements that he and his son are both different and the same. For example, Shiva tells Parvati in the text: "You know that I too can take on six faces. We are speaking only of six elements that he has. My six letters [*bīja-mantra*] also apply to him. Therefore he is only my force. According to his qualities there is no difference between him and me."

of the army of the deity. The nine shaktis asked to remain as the servants of Parvati.

Later, as they strolled past the lake, Shiva ordered Parvati to gather up the six boys in her arms. As she did so they fused into one body with six heads and twelve arms. Shiva, as he had done on the birth of his son, called the child Shanmugan. Parvati loved the boy and fed him milk from her breasts. At this juncture, the six Pleiades maidens left the scene, sent by Shiva to heaven.

The creation is pervaded by dangerous and yet auspicious violence and fragmentation. Shiva himself divides, in a sense, his one head, becoming six. The six sparks take apart the very elements (air, water, earth) of which the phenomenal world is constituted. Parvati breaks her anklet, a common symbol of the vagina, and seems to split into nine forms of herself—her reflections in the gems—each of which is strangely impregnated by Shiva. Each of these nine is herself a more specialized form of Parvati, for each son embodies the particular power of a different gem.

The deity, who is named Shanmugan in the text, is born wholly from the ascetic powers of Shiva. Yet this form created solely from the male principle is unstable: it breaks apart into six. This speaks to the essential composition of Shanmugan: he is replete with potentially destructive power, for he cannot stabilize himself as a unitary being. Parvati, the thwarted mother and here the embodiment of the female principle, reacts with fury and fierceness. She curses the fertility of the wives of the gods and that of the nine shaktis and so turns on herself and inhibits her own nature, enacting in a different mode the destruction of Kama, or erotic desire. The shaktis, in turn, give birth unnaturally to fully grown warriors. The six Pleiades, themselves thwarted mothers who are granted only fosterage, commonly are depicted as fierce beings. Here they also seem to be multiple analogues of Parvati.

But the creation also is auspicious. The sparks of Shiva are transported by basic elements (wind, fire, water), and they become, albeit briefly, the single form of Shanmugan. As O'Flaherty points out: "The seed of the ascetic is never shed in vain, and therefore once shed it must be safely disposed of lest it become destructive instead of creative."[11] And there are other pointers toward the onset of reunification of the fragmented. The nine gems of Parvati's anklet are the *navaratna*, an auspicious configuration, each stone of which is thought to protect against a particular evil or

[11] O'Flaherty, p. 267.

misfortune, and each of which is linked to one of the nine planets (*nava-graha*). The nine warriors who are born of these gems/shaktis, each of whom has the color (and thus the qualities) of his mother, are especially endowed to protect the young deity. In addition, the fierce, virgin Pleiades exit the story once Parvati has fused Shanmugan, while the nine shaktis are reunited with her, as her servants. Thus Parvati begins, albeit on a minor note, to subsume again aspects of her own divided being.[12]

Nonetheless, the chain reaction of division and destruction described above highlights the consequences of the imbalance of Shiva and Shakti. Although Parvati fuses the six forms of Shanmugan, her role, and so that of Shakti on the side of the gods, remains minor. Moreover, her actions are quite controlled by Shiva, who commands her to lift the curse on the shaktis and to assume fosterage of Shanmugan. The outcome of this imbalance is a young deity who is highly differentiated, incomplete, and unintegrated.

The deity played like all children, not aware of his own strength, and so his games and playthings became terrible ones. He piled mountains on top of one another and upended others. Mount Meru he dumped into the sea. He dammed the waters of the Ganges, changed the orbits of the planets, and so forth. A terrified Indra, king of the gods, and others of their number, mistook him for a powerful asura and set forth to slay him. Instead, he killed Indra and many other gods. The Educator of the Heavens told the child: "Indra always intended to serve you, but he could not know that you caused all of this because he could not know you. Therefore he dared to do battle against you, and the Devas [gods] also thought this way. Forgive them and return them to life." The deity did so. He was educable but innocent and unaware of his own being—in other words, he was unreflexive, and in order to mature he had to be taught how his powers were seen by others. His adventures continued. He tamed a wild ram that came forth from a sacrificial fire. He beat and imprisoned the god, Brahma, for not knowing the meaning of the OM, the Pranava Mantra, but later released him at the behest of Shiva. And he slew the asura, Idumban (Iṭumpaṉ) but resurrected him as a devotee.

In exploits such as these, the text describes his shape as that of six heads, or as that of twelve arms, or of both. His form is that of Shanmugan:

[12] Here the unnatural birth of the nine shaktis parallels that of the son of Shiva. Again the offspring of Shiva and Shakti do not harmonize with one another but go to battle together, with the force of Shakti subordinate to that of Shiva.

playful and angry, capricious and fierce, incomplete and somewhat flighty. His being lacks wholeness. He is the product of one sort of asymmetry in the creative relationship of Shiva and Shakti. The asuras are his complement, the result of an obverse asymmetry. Still, his powers in these escapades, although destructive, are not nihilistic. He rends or tames phenomenal form in order to resurrect or to correct and so saves it through his gift of grace.

The Asura Wars: The Assimilation of Shakti and the Encompassment of the Cosmos

Shiva armed Shanmugan with eleven weapons and then with the *vel*. The *vel*, the terrible lance of the deity, is described as nullifying the five basic elements and so as negating all phenomenal illusion. The *vel* (the term coincides with one meaning of "shakti") is identified as the power of Shakti and often as the highest form of Shakti, *jñāna* shakti: the power of discriminating wisdom. According to the tenets of Śaiva Siddhānta, the immutability of Shiva is safeguarded by his acting only indirectly on māyā, the illusion of phenomenal reality, through the mediation of his shakti.[13] Since the *vel* will kill for the deity, something of this mediated relationship is maintained. That he receives the *vel* from his father indicates once more the out-of-balance relationship between Shiva and Shakti. Still, this assimilation of power that is identified with the female principle suggests that some equilibration is occurring in the being of the deity, although the continuing absence of the full integration of male and female principles in his composition points to the operation here of shakti for violent ends.[14]

---

[13] See T. M. P. Mahadevan, "The Idea of God in Śaiva Siddhānta," in *Collected Lectures on Śaiva Siddhānta, 1946–1954* (Annamalainagar: Annamalai University, 1965), p. 4.

[14] Popular understandings modify the asymmetry of Shiva/Shakti, which is so pronounced in the *Kandapurāṇam*, by insisting that the deity received the *vel* from his mother, Parvati. See R. Navaratnam, *Karttikeya—the Divine Child* (Bombay: Bharatiya Vidya Bhavan, 1973), p. 97. One variant I heard in Colombo, of the myth of the competition of Murugan and Ganesha for the fruit of their parents, had it that after Murugan lost and ran off to become an ascetic, his mother tempted him back with the offer of the *vel*. Some worshipers there argued that the deity was unable to defeat the asuras until he received the *vel* from his mother. Others maintained that Parvati had to fast for six days before her son became powerful enough to engage the asuras. This seems in keeping with the inherent powers often attributed to women in Tamil culture. An interesting example of the significance of symmetry in the balance of Shiva/Shakti, and of the creation of the unitary form of the deity, comes from the local myths of the Vaidīśvaran Temple near Mayavaram. Here Parvati is said to have requested the deity to assume a single face, in place of his

After he obtained the *vel*, the deity became wholly destructive for the first time. He entered the fray against Tarakan (Tārakaṉ; Sanskrit Tāraka), the youngest of the three asura brothers. The asura, his vision occluded by illusion (māyā), told the deity that although the gods were the enemies of the antigods, this was not so of Shiva. Therefore the asura could not comprehend why he was battling the son of Shiva. Shanmugan, gradually losing his innocence and increasingly able to discriminate between truth and illusion, between good and evil, sent the *vel* against Tarakan and slaughtered him and Mount Kraunca, his ally and hiding place. This was the first permanent death he caused.[15]

In my interpretation, Shanmugan is becoming more destructive but also more reflexive. According to Kamil Zvelebil's translation, after killing Tarakan the deity states: "The gods and the sages are children of the Supreme One [Shiva]. So are, after all, the Asuras. Hence it is not desirable to destroy them without warning. Let us therefore send a messenger to them. If they oblige and release the immortals [the gods], we shall pardon them; if not, we shall annihilate them."[16] In this utterance the maturing deity simultaneously recognizes similarity and difference and the possibility of their mediation. He perceives contradictions on the level of phenomenal reality but also their resolution on a higher plane. His vision is becoming metaphysical, one that is critically aware of its own consciousness, in comparison, for example, to his earlier instinctive reactions to the attack of Indra. By contrast, Tarakan's previous statement

six. When he did so, she was pleased and rewarded him with the *vel*. See A. K. Chatterjee, *The Cult of Skanda-Kārttikeya in Ancient India* (Calcutta: Punthi Pustak, 1970), p. 71. Unlike the *Kandapurāṇam*, these myths reinsert the positive contributions of Shakti in her own right, rather than rendering this force more peripheral in the triumph of the gods. Thus, in these myths, Parvati prevents the deity from becoming an ascetic like his father, and she induces him to become a holistic being. In both instances, the power of the *vel*, as shakti, is paramount in implying that a holistic being ultimately is preferable.

15 Until he vanquishes the asuras, there is something of the ascetic about the deity. After defeating Tarakan, he embarks on a pilgrimage to places holy to Shiva, in the tradition of the Tamil Shaivite saints and popularized in the puranic literature. Almost all of these sites stress the superiority of Shiva to Shakti. On these saints, see Indira V. Peterson, "Singing of a Place: Pilgrimage as Metaphor and Motif in the Tēvāram Songs of the Tamil Śaivite Saints," *Journal of the American Oriental Society* 102 (1982): 69–90. As well, there are traditions, like that of Sikkal in Tanjore, that the deity did austerities before engaging the asuras. See R. K. Das, *Temples of Tamilnad* (Bombay: Bharatiya Vidya Bhavan, 1964), p. 165. Until he annihilates the asuras, his cosmic career is quite devoid of amorous entanglements. This again points to just how much he is the son of his father, in an ascetic mode, and to the degree to which the influence of the female principle is absent from his being, a lack that is remedied in stages.

16 Kamil Zvelebil, "A Guide to Murukaṉ," *Journal of Tamil Studies* 9 (1976): 1–22, esp. 16.

to Shanmugan before their battle recognizes only polar opposition, similarity or difference, and the impossibility of their confluence on a more encompassing plane.

Shanmugan continued his wars. He killed Singamugan (Ciṅkamukaṉ, Sanskrit Siṃhamukha), the second asura brother, and then fought Surapadman, the eldest and most powerful. Again his form was that of twelve eyes and twelve arms—that of Shanmugan. Before destroying Surapadman, the deity showed to him his own true and eternal form: the cosmos and all of its constituent elements organized hierarchically from the highest (at the head) to the lowest (at the feet) and encompassed in the figuration of a male with one head and two arms, the transcendent male principle that is identified with Purusa, Cosmic Man. In response Surapadman took on a huge, horrific, and cannibalistic form of a thousand arms and a thousand legs. The contrast of this with that of the deity's is instructive. The latter's is systematic and orderly, with a proper place for every existing component part in its encompassment. The former's is chaotic and destructive, devouring, as it were, the very elements that it purports to subsume, in an ultimately nihilistic orgy.

Shanmugan sent the *vel* against this giant and split it in two. Surapadman escaped into the ocean and turned into a gigantic mango tree whose roots grew out of the nether world, whose branches touched the heavens, and whose fruit filled the space of the world. Again Shanmugan hurled the fiery *vel* and split the mango tree in two. Surapadman resumed his usual form, but Shanmugan threw the *vel* a final time, splitting the asura into a cock and a peacock. These birds tried to attack Shanmugan, but he looked at them with love and they were tamed and became good. He took the cock as his pennant and the peacock as his vehicle.

The end of Surapadman bears a certain affinity to the death of Kama, in that each functions as the obverse of the other. The death of the asura cancels the effects of that of Kama, thus bringing to a close the cycle of uncontrolled asymmetry in this myth. In order to explicate the closure of this cycle, it is necessary to specify here how Kama is connected to Surapadman, through the symbolism of tree and mango.

In India, south and north, the symbolism of the tree is associated with procreation and fertility. David Shulman writes that the (cosmic) tree emerges from the nether world which, although a realm of death and disorder, is also a source of fertility.[17] Brenda Beck, in analyzing the

[17] David Shulman, "Murukaṉ, the Mango and the Ekāmbareśvara Śiva: Fragments of a Tamil Creation Myth?" *Indo-Iranian Journal* 21 (1979): 27–40, esp. 27.

homologies between the South Indian Shaiva temple and the human male body, notes that the temple flagpole (i.e., tree) corresponds to the genital region.[18] Parvati sometimes nourishes a tree in place of the infant Skanda.[19] Elsewhere she proclaims that "a tree is equivalent to ten sons."[20] And, in one version of the Skanda myth, Parvati curses all the gods to become barren trees.[21] In another, the god Vishnu creates a mango tree to shade his sister, the enticing and seductive Devi (Uma/Parvati), as she does austerities to attract Shiva. Shiva incinerates both Devi and mango tree with the power of his third eye.[22] In Sri Lanka, pervaded by the influence of Southern Hinduism, the mango is an explicit symbol of the vagina, and the goddess Pattini, the Sinhala analogue of Uma/Parvati, is born of a mango.[23] The symbol of the tree seems to connote the power and florescence of eroticism and fertility.

In these and other myths, Kama frequently is associated with the symbolism of the tree. According to O'Flaherty, Kama is often "pictured as a tree whose sprouts are women."[24] And, in trying to escape the wrath of Shiva, Kama at times either hides behind a tree or becomes one, but to no avail, for the tree/Kama is ignited by Shiva.[25] In this configuration, the symbolism of Kama is that of erotic desire and fertility; the symbolism of the tree is associated with desire, fertility, and disorder; Kama is associated with a tree that is burned by Shiva, as is Kama himself; sometimes this tree is a mango; and the mango may be a symbol of the female and of seductiveness—that is, a representation of shakti.

The fiery rending of the mango tree by the *vel* brings to a close the reverberating cycle of asymmetry that was intensified by Shiva's incineration of Kama. The destruction of Kama signifies the tearing away of normal fertility from the Godhead, while the destruction of mango tree ~ Surapadman signifies the restoration of these forces. The death of Kama signifies the estrangement of Shiva from Shakti, of asceticism from normal fertility, and the disruption of the karmic cycle. In part, the maturation of the deity, from an unstable, self-divided, and incomplete being into an

[18] Brenda E. F. Beck, "The Goddess and the Demon: A Local South Indian Festival and Its Wider Context," *Puruṣārtha* 5 (1981): 83–136, pp. 120–22.

[19] Shulman, "Murukaṉ," p. 35; O'Flaherty (n. 7 above), p. 160.

[20] Shulman, "Murukaṉ," p. 35.

[21] O'Flaherty, p. 160.

[22] Shulman, "Murukaṉ," p. 33.

[23] Gananath Obeyesekere, *The Cult of the Goddess Pattini* (Chicago: University of Chicago Press, 1984), pp. 456, 458, 470.

[24] O'Flaherty, p. 159.

[25] Shulman, "Murukaṉ," p. 33; O'Flaherty, p. 160.

encompassing and holistic one is the process of recouping and assimilating the power of the female principle and the placing of this under the control of that of the male. To recapitulate: the six forms of Shanmugan are fused by Parvati, at the command of Shiva, and Shanmugan obtains the *vel* from Shiva. The *vel* is considered homologous with the third eye of Shiva. This weapon acts under the control of Shanmugan, and is identified also with *para*-shakti or *jñāna* shakti, the power of discriminating wisdom and the highest form of shakti. By contrast, Surapadman is the offspring of Māyā, of lower forms of shakti (to be discussed further on) that have magnified and spiralled out of control and that are not under the direction of the male principle.

I have argued that Kama's death signifies the cessation of erotic desire and normal fertility in the world of saṃsāra. On the side of the asuras, the monstrous growth of the mango tree, which fills all space, stifles all growth, and freezes all movement, is a perverted cosmic tree that grips the cosmos in stasis. The gigantic mango tree signifies sterility—the uncontrolled growth of desire without fertility. It is the extreme complement of the weakness and futility of Kama in relation to Shiva, on the side of the gods. Thus the *vel*, or higher-order shakti in the service of the deity, rends the mango tree, or lower-order shakti that has spiralled out of control on the side of the asuras. But the mango tree ~ Surapadman is not utterly annihilated. Instead, it is transformed into cock and peacock. Like the cosmic tree, "in India both cock and peacock are linked with death and the nether world"[26] and so to connotations of disorder and fertility. In my interpretation, that this cosmic tree is a mango strengthens these connotations of sexuality and fertility. But there is here a singular contrast to Shiva's incineration of Kama. Shiva totally nullifies Kama's power to influence him, and so he negates this side of himself as the cosmic being. Shanmugan ~ Murugan assimilates cock and peacock, and so, by extension, he also incorporates within his being those lower orders of shakti that had been torn from the Godhead, to make their own way in the cosmos. Later on in the myth, Murugan marries two consorts, two shaktis, whom he controls. These should be understood as objectifications of the lower-order powers of shakti that he assimilated to his being in the defeat of Surapadman.

I surmise that the splitting of Surapadman was the apex of the destructive expressions of the deity as Shanmugan. Surapadman was neither utterly

[26] Shulman, "Murukaṉ," p. 32.

annihilated nor given the grace of release by the deity. Instead, his state became one of static fragmentation in this life, his soul perhaps removed from the karmic cycle of death and rebirth. In a sense, his fate paralleled that destined for the cosmos under the rule of the asuras. Therefore this inversion underlined the creative destructiveness of the deity, for through these actions he ushered in a new age, the Age of Kali, the beginning of a new cycle of cosmic time. Yet the fate of Surapadman may be understood also as the ultimate sacrifice on the part of the devotee—to remain permanently in this world, fragmented and suspended, in the service of the deity. This state is reminiscent of that which Friedhelm Hardy calls the "torn personality" of the devotee, the bhakta, who is condemned to suffer the longing for union and the agony of separation from the deity.[27]

That the cycle of asymmetry and fragmentation closes with the rending of Surapadman is supported by what follows. After the slaying, the *vel* flew to the Ganges and dipped in its waters. This seems to be an act of purification that ends the pollution of rage, battle, and blood. If this interpretation is valid, then other related myths of the deity support the view that we have a shift from the furious multiform of Shanmugan to a more holistic, reflexive, and encompassing being. According to the myths of Sikkal, in Tanjore, Surapadman was a Brahman. In killing him the deity committed the sin of brahmanicide. His body lost its lovely shape and turned pitch black in color. At Sikkal a self-created lingam (i.e., Shiva) emerged from a pond of milk. The deity worshiped the lingam, bathed in the milky pond, was released from his sin, and regained his normal coloration and form.[28] And, writing of Tiruchendur (Tiruccĕntūr), where the deity is said to have destroyed Surapadman, Fred Clothey notes: "The icon of the god … is not that of a warrior but of a penitent, for it was at this spot, local tradition maintains, that the god expressed his regret for having engaged in violence."[29]

The text itself seems not to give vent to such sentiments or consequences. But as this canto closes, Murugan, with one head and two arms, seated on his throne, is worshiped by the assembled beings of the cosmos as the dawn breaks. He appears to be treated as a completed and holistic deity, transcendent and encompassing of previously imbalanced forces.

---

[27] Friedhelm Hardy, *Viraha-Bhakti: The Early History of Kṛṣṇa Devotion in South India* (Delhi: Oxford University Press, 1985).

[28] Das (n. 15 above), p. 165; see also Navaratnam (n. 14 above), p. 50.

[29] Fred Clothey, "Pilgrimage Centers in the Tamil Cultus of Murukaṉ," *Journal of the American Academy of Religion* 40 (1972): 79–95, esp. 82.

The Asuras: The Asymmetry of Shakti

Thus far the force of Shakti has remained quite controlled and minimized by Shiva. Parvati is almost peripheral to the existence of her son. Shiva's own composition initially is imbalanced and lacks integration. The tale of the asuras exemplifies the obverse condition: the consequences of Shakti uncontrolled and of Shiva (as the male principle) minimized. The asuras are stable in a static sense: their condition produces atrophy, entropy, and chaos. By contrast, Shanmugan is dynamic and creative, but not stable. His slaying of the asuras is generative: it ends stasis and reintroduces the karmic processes of death and rebirth. Therefore the story of the asuras should be read as an inversion of that of the deity. In particular, this story revolves around the dominance of the principles of Shiva or Shakti and the recreation of symmetry through the mutual obliteration of both sides of the inversion. The deity as Shanmugan is made aware of his regulatory aspect, as the being I call Murugan, and comes to transcend and so to encompass his less reflexive self.

The Birth of the Asuras: Seduction and Illusion

The tale of the asuras stresses the dominance of Shakti. Māyā, the daughter of the lord of the asuras, is an inversion of Parvati. She is the embodiment of Shakti rampant and untrammeled, in comparison to the stunted version that is Parvati. After her birth, the daughter of the asura lord studied the science of māyā, of wealth and illusions. She was so adept that her teacher gave her the name of Māyā. He told her that the god, Vishnu, had killed many asuras and had reduced the powers of the asura lord. Only she could recoup their waning fortunes. She was to transform herself into a beautiful and desirable woman and seduce the sage, Kashyapa. They would have many asura offspring to whom she would teach the rules of sacrifice and asceticism.

Kashyapa and Māyā mated and created offspring, three sons and a daughter, all the while going through changes of shape. Māyā then discarded Kashyapa. Her role was crucial to the creation of the asura children, while Parvati's was peripheral to that of Shanmugan. Illusion, as both cause and effect,[30] was basic to the accomplishments of Māyā. By

[30] Prabhu Dutt Shastri, *The Doctrine of Māyā in the Philosophy of the Vedānta* (London: Luzac, 1911), p. 31.

contrast, Shanmugan was created by the third eye of Shiva, the eye of wisdom that pierced all illusions.

Like Māyā, Shiva and Parvati commonly were said to have four offspring: Ganesha, Virabhadra, Bhairava, and the youngest, Subrahmanya (i.e., Shanmugan ~ Murugan). The children of Shiva and Parvati, and those of Māyā and Kashyapa, have strong affinities and are complementary in their attributes. Virabhadra and Bhairava are fully the creations of Shiva. Virabhadra was born from the sweat of Shiva or sprang from a lock of his father's hair—in either instance, to destroy Daksha, the father of Parvati.[31] Bhairava was born of the blood of Shiva, and his image often stands guard at the portals of Shaiva temples.[32] In South India both are thought of as wholly grim and fierce beings. On the side of the asuras they are paralleled in their attributes by the two younger brothers, Singamugan and Tarakan.

Ganesha is wholly the creation of his mother, Parvati. According to myths of his birth, she made him of her sweat, of her skin rubbings, of the water in which she bathed after making love to Shiva, or of turmeric. In all instances he was created to protect the entry to her chamber. He prevented the access of Shiva, who cut off his head. After the entreaties of Parvati, Shiva replaced the head with that of an elephant. He was made the first son of Shiva (hence his name, Pillaiyar, the son) and the leader of the Shaiva demigods, the *gaṇas* (and so his name, Gaṇapati).[33] He is the Lord of Categories who inserts or who extracts obstacles. He is thought of as somewhat effeminate or asexual, and he often is depicted as a bachelor who sits on the river bank, watching the girls bathe and waiting for one who resembles his mother. On the side of the asuras, his attributes contrast with those of the daughter, Ajamuki (Acamuki; Sanskrit Ajamukhī). She is sexually rampant and unbridled, for she rapes all men she meets, and, cannibalistic, she is destroyer of categories.

The major protagonists, Shanmugan and Surapadman, have a complicated relationship. They are almost mirror images, although the being of the former evolves and develops while that of the latter does not. Both are the "sons" of Shiva. Shanmugan is his "natural" son, on whom he bestows the power to annihilate the antigods. Surapadman is the devotee par

---

[31] G. Jouveau-Dubreuil, *Iconography of Southern India* (Paris: Librairie Orientaliste Paul Geuthner, 1937), p. 48; H. Krishna Sastri, *South Indian Images of Gods and Goddesses* (Madras: Madras Government Press, 1916), p. 155.

[32] Krishna Sastri, p. 151.

[33] Ibid., p. 165.

excellence, to whom he grants the power to freeze karmic processes and to overturn the dharmic order.[34] Both have six heads,[35] and, as Zvelebil points out, they use similar weapons in their climactic battle.[36] Yet the weapons of the asura are imitations of those of the deity and as such are less focused and less controlled than the deity's. Given their origins, their complementarity is complete: Shanmugan is the creation of the extreme inner heat of asceticism; Surapadman, of the unbridled heat of sexual abandon, of woman uncontrolled by the bond of marriage. These oppositions, embedded in the composition of Shiva, are brought to a heightened pitch in the characters of the combatants. According to the text, the signal distinction between them is their respective ability to perceive true reality. Thus the true (i.e., highest) being of Shanmugan is not his furious multiform of six heads and twelve arms but that of Cosmic Man with one head and two arms, subsuming the ordered cosmos. But the true being of Surapadman is indeed a multiform of one thousand heads that depends on illusion for its existence and that cannot transcend this form without further disrupting the cosmos.

The Education of the Asuras: The Value of Materialist Reality

Kashyapa instructed his children in the tenets of Śaiva Siddhānta and in the kind of devotionalism, or bhakti, that Hardy calls "intellectual," or bhakti-yoga.[37] The preoccupation of the wise, he told them, is to study God or essence (*pati*), the individual being or soul (*pacu*; Sanskrit *paśu*), and their connection or passion (*pācam*; Sanskrit *pāśam*) that relates and that separates them. Only Shiva is *pati*. The numberless people in the world are the objects of illusion, luxury, and pride (the *malas*), and, because of these sins of the world of saṃsāra, they are born and die. Knowledge, bravery, wealth, and power, Kashyapa added, also are transitory. Only righteousness is stable and provides happiness in this life and

[34] Still, Shanmugan has the advantage. As the text makes clear, he and his father are one and the same, are different, and are interchangeable. Tamil understandings of kinship relationships emphasize the interchangeability of father and son and, in tandem, the merging of godhead and child. See Brenda E. F. Beck, "The Kin Nucleus in Tamil Folklore," in *Kinship and History in South Asia*, ed. T. R. Trautman (Ann Arbor: University of Michigan Center for South and Southeast Asian Studies, 1974), pp. 15–16.

[35] Fred Clothey, "Skanda-Ṣaṣṭi: A Festival in Tamil India," *History of Religions* 8 (1969): 236–59, esp. 253–54.

[36] Zvelebil (n. 16 above), p. 21.

[37] Hardy (n. 27 above), p. 291.

in that to come. Māyā laughed at these words and mocked the teachings of Kashyapa. Knowledge and wealth, she said, were important; the material world was the only reality.

Her sons, avid devotees of Shiva, performed elaborate and severe austerities for a lengthy period, but to no avail. Surapadman sliced off pieces of his living flesh and tossed them into the sacrificial fire, yet Shiva did not respond. Then Surapadman, in ultimate self-sacrifice, threw himself into the flames and was consumed. Shiva resurrected him and granted all his desires: universes to rule, a lion mount, weapons to conquer the gods, and an eternal, invulnerable body. Only his own power, said Shiva, could conquer the asuras.

The brothers returned to Kashyapa, who washed his hands of them and sent them to study with his guru. These teachings continued the materialist themes propounded by Māyā: they denied karmic processes, dharmic order, and the distinction between truth and mistaken perception (ahaṃkāra). Robert Perinbanayagam argues, I think correctly, that there is a utopian vision in the position of the asuras, one that rejects the hierarchy and suffering of life on earth.[38] However, this position rests on the values of materialism and egoism, and it denies any significance to the realization of the self or to the problematic harmonization of soul and Godhead. This position denies the spiritual development of the soul, whether through karmic processes central to Śaiva Siddhānta or through bhakti devotionalism. Therefore, in terms of these ontologies it is an antireflexive position, one that promotes simple reflectivity or the ultimate value of sentient consciousness.

The asura brothers battled and conquered the gods. Surapadman, the supreme devotee, received the blessings of Shiva. In the new asura order, the moon was to remain whole; the sun, to be heatless and to shine at night; Agni (the god of fire) to turn cold when touched by a servant of the asura; and so forth. Death was abolished; the gods were forced to become fishermen (i.e., killers of fish), and they ceased to follow the laws of the dharma they were required to keep.

The asuras reconstitute the cosmos as an unstable place, as an aggregation of elements rather than as a system.[39] But this instability is not the

---

[38] Perinbanayagam (n. 9 above), p. 68. Among the most noteworthy of the guru's teachings were that what one did in this life had no bearing on the life to come and that everyone who existed was equal—there were no superiors or subordinates.

[39] Bruce Kapferer, *A Celebration of Demons: Exorcism and the Aesthetics of Healing in Sri Lanka* (Bloomington: Indiana University Press, 1983).

product of flux but, rather, of stasis—of an entity devoid of dynamism and generative forces, in which karmic processes grind to a halt and are superseded by those of entropy. This state mirrors the constitution of the asuras themselves. Their being too is static, for it neither evolves nor transforms itself into higher-order levels of perception, discrimination, and encompassment. Shanmugan also is unstable, but his instability is that of action and innovation, of the creative source in the process of evolution, whose intervention restores dharmic stability.

The Incorporation of Evil

The text opposes Shiva to Shakti on each of the respective sides of the gods and the asuras, and between gods and asuras. On the side of the gods, Shakti (the female principle) is wholly a function of Shiva (the male principle), and this asymmetry creates the skewed Shanmugan. On the side of the asuras, the male principle (Shiva) is solely a function of the female principle (Shakti). Māyā is supreme, and her mate is deluded and peripheral. This produces the warped asura children. In either instance, the result is destructive: one more controlled and creative, the other more imitative, static, and therefore chaotic. Indeed, at one point in his description of the climactic battle, the commentator says that the struggle was so equal that the Goddess of Victory did not know to whom to award the triumph.

In terms both of Śaiva Siddhāntin theology and of popular Tamil beliefs, this opposition between Shiva and Shakti does not require any complete or ultimate resolution, for this would produce a static cosmos. Instead, the desired condition is one of dynamic tension that produces the ongoing blending of symmetry and asymmetry in relationships between these two vital principles. This tension enables the activation of each and the conjunction of both. Thus, through his conquest of the asuras, Shanmugan, the product of Shiva, incorporates more fully the functions of shakti that they embody.[40]

Entering battle, Shanmugan still is a poorly integrated and unstable composition. The character who emerges triumphant is on his way to becoming a more encompassing and holistic being. Surapadman, entering the fray, is the essence of chaos (as a product of stasis in this cosmology).

[40] Shanmugan kills the asura brothers in order of ascending age. This is an effective dramatic device that builds climax. But it also may signify the progressive and additive assimilation of the increments of evil into his own being.

He emerges a permanent fragmentation, the embodiment of stasis, of frozen immobility, incapable of reintegration. Thus his fate is precisely that which he had decreed for the cosmos. He has become the pennant and the vehicle of the deity, completely under the control of the latter, and he can act only at the behest of his master. This is one version of an asymmetry between the principles of Shiva and Shakti. The force of illusion, controlled by the shakti of the divine, becomes incorporated into the being of the triumphant deity. The battle between Shanmugan and the asuras is a confrontation between Shiva and Shakti, in which all of the latter comes fully under the rule of the former. Other myths of Shiva also fit well with this text.[41]

Through their confrontation, the inversion of "good" (active in the pursuit of regeneration) and "evil" (active in the pursuit of stasis) cancel one another. I would also argue that the deity fights and surmounts an aspect of himself that he is either unaware of or lacks. In battling Surapadman, Shanmugan wars also with his own shadow side, the asura as his mirrored self, the potential for devolution and fragmentation that are part of his being as the divine encompassment. His trial is a conquest over self that has the appearance, and hence the illusion, of the Other (i.e., of Surapadman). As the Other disintegrates, the fuller self, that of self-realization, comes into being. By conquering evil, as the Other, Shanmugan makes it integral to himself, to his knowledge of self. Since, like his father, he is identified with the cosmos and encompasses it, he must have the self-constitution to transcend and thus order the cosmos. To accomplish this, he must know his own being completely—in other words, he must be fully reflexive. By destroying evil he comes to know this quality, to subsume it, and so to transcend this as part of himself. He is no longer the playfully rambunctious child whom the god, Indra, mistook for an asura. And, in metaphysical terms, it hardly is coincidental that he is thought to have committed brahmanicide in his rending of Surapadman. In this terrible act he experiences evil fully and makes it part of his own being. He thereby makes a place for it within the cosmos identified with him. He awakens to evil, becomes conscious of it, as the asuras could never do. For Shanmugan the immediate consequence is the highly "heated" condition of extreme impurity. Therefore, according to the text, the *vel* washed itself in the Ganges, while, according to other myths, the deity did more extreme penance to cool his being.

---

[41] See Fuller (n. 8 above).

At this stage of his evolution the deity is constituted such that the principles of Shiva and Shakti are related to one another within himself. Ontologically he is in the position of his father, Shiva, before the latter created Uma/Parvati and so objectified the force of shakti. This brought Shakti under Shiva's control yet freed this force to function more independently in the phenomenal world. This in turn freed Shiva to pursue asceticism when he so chose. In his ontogenesis, Shanmugan ~ Murugan has reached the pinnacle of evolutionary development, that of holism and encompassment. He has closed the cycle of fragmentation and so has replicated the origins of the cosmos. But the next stage, the devolution of divine power to define and to differentiate the phenomenal world, has yet to occur.

Therefore Shanmugan ~ Murugan objectifies the power of shakti in order to free this force to operate on and in the world and so to shape and order reality in all of its finer gradations and manifestations. In so doing he goes beyond the dualism of Shiva and Shakti, or of Shiva and Parvati, for he brings into existence the triad of the deity and his two wives. Through his marriages, the deity continues to be re-created and reconstituted in a more balanced but necessarily dynamic condition, as is the cosmos with which he is identified. In other words, by objectifying the force of Shakti, the deity opens the way to becoming both a transcendent and an immanent god.

From Encompassment to Immanence

Two daughters of Vishnu, Amutavaḷḷi and Sundaravaḷḷi, fell in love with Murugan and desired to wed him.[42] He replied that Amutavaḷḷi would be reborn the daughter of Indra (she received the name Devayanai), that Sundaravaḷḷi would be reborn the daughter of the sage, Shivamuni, and that he would marry both. After he slew the asuras, the deity married Devayanai, and they lived royally and quietly in the abode of the gods.

In the meantime, in the hill country of Tamil Nadu, Shivamuni was aroused from his austerities by the sight of a lovely doe, whom he impregnated with his glance.[43] She birthed a daughter but abandoned the infant

---

42 According to Brenda E. F. Beck, "A Praise-Poem for Murugan," *Journal of South Asian Literature* 11 (1975): 95–116, esp. 108, the *Kandapurāṇam* states that these two were, respectively, older and younger sisters in a previous birth.

43 The last chapter of the text, the "Vaḷḷiyammai Tirumaṇappaṭalam," states that Shivamuni is a manifestation of Vishnu, and the doe, that of his wife, Lakshmi.

among wild tubers. Simple hunters discovered the baby, named her Vaḷḷi ("creeper"), and raised her as their own. Told by the sage, Narada, of her beauty and devotion, Murugan left Devayanai in the abode of the gods and sought Vaḷḷi. He was smitten by her loveliness. Here the text describes him again as multiform—that is, as Shanmugan. Gone was his quiescence with Devayanai. He desperately pursued the love of Vaḷḷi and became passionate and ecstatic. He went through shape changes, used trickery, and threatened virtual self-castration in order to win her heart. These themes of illusion, deceit, and disintegration are quite alien to his relationship with Devayanai. Theirs is characterized by *kaṟpu*. George Hart, writing of the ancient Tamils, translates this as female "chastity" that connotes "a sort of asceticism, restraining all impulses which were in any way immodest." The woman who possessed this quality produced "domestic peace and light" and protected the males of her household. Beck describes this quality as the chastity and love that arise from marriage and from formal obligations and relationships.[44]

To court Vaḷḷi, the deity took the form of a hunter, a tree, a hunter once more, and then that of an old Shaivite sage who proposed to her but was rebuffed. He called on his brother, Ganesha, to take the shape of a wild elephant, whose sudden, ferocious presence frightened the maiden into the arms of the deity. Terrified, she acquiesced under this duress to wed the old man. He revealed his identity to her: his multiform of six heads and twelve arms, passionate and furious. The clandestine courtship continued. Vaḷḷi pined for her lord to the degree that her family thought her possessed by an evil spirit. He, driven to the point of madness, threatened to ride in the village on a *maṭal* hobbyhorse, made of the jagged stems of palmyra leaves. According to Hart, this was a poetic convention whereby a lover, by shedding his blood in public, shamed his beloved into acceptance.[45] Hardy adds that to ride the hobbyhorse was "to make an utter fool of oneself, to be laughed at and gossiped about by the villagers."[46] Later the couple eloped and were chased by Vaḷḷi's father and others of the village. As the multiform Shanmugan, the deity killed the harassers. Yet he revived them after the sage Narada explained to him that such destruction was not correct behavior. A proper wedding followed, and the tāli, the Tamil wedding necklace that binds the bride to her spouse, was

---

[44] George L. Hart III, "Woman and the Sacred in Ancient Tamilnad," *Journal of Asian Studies* 32 (1973): 233–50, esp. 236–37; Beck, "A Praise-Poem for Murugan," p. 107.

[45] Hart, p. 244.

[46] Hardy, p. 396.

tied. Shiva, Parvati, and other gods attended and gave their blessings. As Shulman discusses in detail, the whole of the courtship episode is permeated with erotic metaphors of desire and laden with the wild longing of the lovers.[47]

Shakti and the Devolution of the Divine

The deity's relationship with Vaḷḷi is that of *kaḷavu*, defined by Beck as love prior to marriage and as the attainment of a higher condition of existence without regard for the strictures of social norms. Vaḷḷi, argues Beck, represents the worship of the divine through "ecstasy and self-abandonment. Self-abandonment, in Hinduism, is associated with lack of control, and consequently with danger and defilement in general." By contrast, Devayanai represents "the conventional, controlled, ritualized approach to worship."[48]

Shulman characterizes the courtship of Vaḷḷi as a myth of bhakti, but one in which the divine pursues the soul of the devotee in order to achieve union, rather than the conventional reverse of this.[49] And, according to J. L. Brockington, the Vaishnava bhakti of Ramanuja implies that the "deity needs the soul."[50] One may argue that, in the hypothetical extreme case, just as the devotee, the bhakta, can lose his sense of self-awareness and sense of being in that of the divine, so too can the god lose himself in the being of the devotee. In her analysis of a present-day popular pamphlet that tells of the courtship, Beck notes that the deity is referred to as a "drunken reveller" and as one who is furious and possessed, trembling with confused mind, and full of deceit.[51] The character of the being who pursues Vaḷḷi is highlighted further in various tales that blame Ganesha's elephant head on the feverish condition of the deity. In these stories,

---

47 Shulman, *Tamil Temple Myths* (n. 3 above), pp. 277–82.

48 Beck, "A Praise-Poem for Murugan," p. 108.

49 Shulman, *Tamil Temple Myths*, p. 280.

50 J. L. Brockington, *The Sacred Thread: Hinduism in Its Continuity and Diversity* (Edinburgh: Edinburgh University Press, 1981), p. 136.

51 Brenda E. F. Beck, "The Courtship of Valli and Murugan: Some Parallels with the Rādhā-Krishna Story," in Hawley and Wulff, eds. (n. 9 above), p. 272. In this version, at the outset of the courtship the deity is delineated through metaphors that connote heat, while those of Vaḷḷi connote coolness. As their relationship develops, the metaphors for the deity become cooler (esp. after he reveals his divinity), while those of Vaḷḷi increase in heat as she softens and loses her heart. Once again it is the quality and conditions of relationship that mold their respective characters.

Shanmugan behaves in a deluded and irresponsible manner.[52] In love, as in war, the deity in this condition is highly active and imbalanced.

The text states that, after a period among the hunters, the deity and Vaḷḷi joined Devayanai in the abode of the gods. The purana closes by saying that the trio lived together in harmony and compares them to the three eyes of Shiva. Some texts compare the triad to three shaktis or identify the two consorts with two shaktis. Thus Devayanai is identified with the force of *kriyā* shakti: the power of motivation, of works and action, and of maintenance. Vaḷḷi is identified with the force of *icchā* shakti: the power of desire, will, and delusion that is passionate and disintegrative.[53] The deity himself, or the *vel*, which may be worshiped in place of him, is identified (like the third eye of Shiva) with *jñāna* shakti: the transcendent power of true perception and of discriminating wisdom that encompasses the other two and so equilibrates them and the cosmos. In this schema, *icchā* shakti is nearest to phenomenal reality, *kriyā* shakti is further from the material world, and *jñāna* shakti is the most distant and abstract. The shaktis are ordered hierarchically, as is the constitution of the deity as he moves among them. On the one hand, he subsumes the shaktis, but, on the other, he is influenced, and perhaps reconstituted, by them. This will be discussed and qualified further on. For the moment, it is necessary to note that, in the theology of Śaiva Siddhānta, these identities are fraught with implications for the logic that informs the composition of the deity and, so, of the cosmos. They require some brief discussion here.

In the ontological schema of Shaivism, the most abstract conception of shakti is that of *prakṛti māyā*, the power of consciousness, thought of as female, that activates the abstract and inert male principle of puruṣa and so brings the world of phenomenal reality into existence. From prakṛti māyā there evolved a further twenty-four evolutes, or categories (*tattvas*), of increasingly greater specialization in giving the world its shape and substance. Among the last of these are the trio that the ontology of the older Sankhyan philosophy calls guṇas, or humors, and that Shaivism sometimes refers to as the "sons of māyā." Here māyā is one of the three *malas*, sins or bonds of the spirit, that the soul must sever to attain union with the divine (the two others being *āṇava*, or ignorance, and karma). The three guṇas, which together produce the lowest levels of phenomenal reality—perceptions of material nature—are called sattva,

[52] Richard L. Spittel, *Far-off Things* (Colombo: Colombo Apothecaries, 1933), p. 285.

[53] See Shulman, *Tamil Temple Myths*, p. 282; Zvelebil (n. 16 above), p. 10.

rajas, and tamas. The quality of sattva produces the organs of knowledge, that of rajas the organs of action, and that of tamas the subtle essences and the five basic elements.[54] These three qualities are not particulates or isolates in the world of existence. Instead, all three are present in all beings, whether gods or humans, and the given, although shifting, balance of the three in relation to one another determines in large measure one's condition of being or character. Thus the sattvik quality is identified with the truer perception of phenomenal reality, the rajasik quality with the attachment to reality, and the tamasik quality with torpor or ensnarement in mundane reality.

The three guṇas summarize the powers of māyā as these are manifest in phenomenal existence. However, as noted before, the Shaivite conception of Godhead does not act directly on the phenomenal world through māyā. Instead, this is done through the mediation of shakti. In the Agamic or orthodox tradition of ritual observance, the three guṇas are basic to the three shaktis.[55] In other words, these three shaktis, each a quality of power or force, resonate with the three *guṇas*, each a quality of matter: *jñāna* shakti with the sattva guṇa, *kriyā* shakti with the rajas guṇa, and *icchā* shakti with the tamas guṇa. The finer gradations between kinds of shakti enable God to veil his supreme being and, therefore, to enter into, to have compassion for, and to act on the most mundane of beings constituted through a balance of the gunas.[56]

Iconography, Reflexivity, and Shakti

This connection between shakti and guṇa encourages another look at the iconography of the deity and at the implications of this for the logic of his evolution and devolution. The language of iconographers of South India is often couched in Sankhyan terminology since the images discussed are classic representations. Therefore, the relevant discourse here is that on the relation of the guṇas to their expression in figuration, rather than that of the shaktis, which is given more prominence in Shaivite metaphysics.

---

[54] See Mahadevan (n. 13 above), pp. 6–7.

[55] See Teun Goudriaan, *Māyā Divine and Human* (Delhi: Motilal Banarsidass, 1978), p. 168.

[56] See C. Mackenzie Brown, "Theology of Rādhā," in Hawley and Wulff, eds., pp. 67–68. Although his comments are directed at the Vaishnava tradition of bhakti, in this context they apply as well to that of Shaivism.

Gopinatha Rao, an astute and accurate observer of Hindu iconography, comments that "sculpturally it may be said, the number of hands in an image represents the number of attributes belonging to the deity.... The larger the number of hands, the more numerous are the attributes conceived; and the image with many hands is considered to be tamasic in character... the less the number of hands, the less numerous are the gunas conceived; and the image may thus be made to approach as nearly as possible the description of the attributeless Brahman."[57] More specifically, with regard to Subrahmanya (or Shanmugan ~ Murugan), Rao writes that an image with only two arms is said to be sattvik, one with four arms is said to be rajasik, and one with more than four arms is said to be tamasik in character.[58] These stylistic tendencies are echoed in the thrust of connotation—if not in the exactness of description—of comments by other iconographers. H. Krishna Sastri writes that a text, the *Śilpasaṅgraha*, mentions three varieties of Virabhadra who are sattvik, rajasik, and tamasik and who have respectively two, four, and eight arms. In a similar vein, he describes three aspects of the Shaivite temple guardian, Kshetrapala. Teun Goudriaan mentions that the Jain goddess Ambika (i.e., female, mother) in her sattvika form is seated and has two arms, while in her rajasa form she stands upright and has eight arms. More generally, G. Jouveau-Dubreuil notes that "Brahminical deities have only four arms when they are in repose.... The number of arms is only more than four when the deity is represented in movement."[59]

A clear statement of ethnographic detail appears in Beck's discussion of a trio of local goddesses who may be thought of as a set of sisters. The highest form is that of Parvati: she is depicted with four arms, two of which are used to bless devotees, and two of which are empty. Both of Parvati's legs touch the ground, and she wears a mild expression. The lowest form is that of Bhadrakali: she has eight arms, all of which hold weapons, and she wears a fierce expression. Bhadrakali's left leg is bent in a dance posture, and beneath her left foot lies a human male warrior holding a sword. Her right leg is raised. Between these two is Mariyamman, "a kind of compromise between these fierce and mild forms." Mariyamman is seated, but with her right leg dangling. She has four arms, all of which grasp objects.

---

57 Gopinatha Rao, *Elements of Hindu Iconography* (Madras: Law Printing House, 1914), 1, pt. 1:27.

58 Ibid., 2, pt. 2:425.

59 Krishna Sastri (n. 31 above), p. 159; Goudriaan, p. 174; Jouveau-Dubreuil (n. 31 above), p. 128.

Two of these are ritual objects (like Parvati), and two are weapons (like Bhadrakali). Mariyamman and Bhadrakali are "descending forms" of the goddess during her time on earth, while Mariyamman is "more active than Parvati but less active than Bhadra Kali."[60] Sheryl Daniel's statement that "the more differentiated are emanations of the less differentiated" sums up this general picture well.[61]

Though brief and fragmentary, these details mesh with the idea that when a god is constituted by a preponderance of sattvik qualities, his mode is one of ascendancy, of aloofness and distance from the phenomenal world, and of transcendence. When he is constituted of a preponderance of rajasik qualities, his mode of being is one of action in the service of maintenance, of guarding the formal regulation of the cosmos. But when he is constituted primarily of tamasik qualities, his mode of being is one of devolution, of immanence in the phenomenal world. These conditions of being are expressed through iconography, whether in the imagery of the icon or in the metaphor of the text. Thus the multiform Shanmugan is tamasik in character and is driven by the force of *icchā* shakti. In accordance with this line of reasoning, the divine in the sattvik condition is informed by the wisdom and perception of *jñāna* shakti, while in the rajasik condition he is influenced strongly by *kriyā* shakti in his regulation of the phenomenal world. He is formal in his relationship to his devotees and is worshiped as such, rather than through passionate bhakti that is more in tune with the divine in a tamasik condition. Indeed, when the deity is not the multiform Shanmugan, he becomes the more uniform Murugan, with a holistic shape of one head and two arms. In this condition he neither goes to war nor falls in love, and his dwelling is farther from the devotee.

At least one region of Tamil culture has institutionalized the kinds of distinctions explicated above. According to Bryan Pfaffenberger, the dominant Veḷḷāḷar caste of the Jaffna Peninsula in Sri Lanka distinguishes markedly between temples dedicated to Skanda and to Murugan. All agree that these are the same deity, but each is worshiped as a distinctive being. Skanda is orthodox and Brahminical, and he should be worshiped

60 Beck, "The Goddess and the Demon" (n. 18 above), p. 101.

61 Sheryl B. Daniel, "Marriage in Tamil Culture: The Problem of Conflicting Models," in *The Powers of Tamil Women*, ed. Susan S. Wadley (Syracuse, N.Y.: Syracuse University, Maxwell School of Citizenship and Public Affairs, 1980), p. 86. See also Wendy Doniger O'Flaherty, *Women, Androgynes, and Other Mythical Beasts* (Chicago: University of Chicago Press, 1980), p. 91.

only in temples that follow the correct Agamic ritual procedures and as an "invited" deity: one who is transcendental, regal, aloof, fearful of pollution, and relatively unresponsive to worshipers. Murugan embodies the spirit of bhakti. He is worshiped as a "chosen" or "personal" deity who will give any boon to his devotee. Skanda is the proper ruler of the cosmos, while the only true shrine of Murugan is in the jungle at Kataragama, in the distant southeastern corner of the island. There he is immanent as a "self-born" god: one who through his love and compassion chose of his own volition to reveal himself. Jaffna Veḷḷāḷars consider that Murugan's propensity to reveal himself to women, fisherfolk, and wild Veddahs shows just how threatening he is to cosmic order.[62] But Jaffna Veḷḷāḷars apparently do not emphasize the transformational properties of the multiform ~ uniform deity.

Earlier I argued that Shanmugan's conquest of evil was a battle for the consciousness of self-awareness. He came to know the furiousness of evil by participating in its act. Himself the precarious product of the principle of Shiva separated from Shakti, his victory incorporated into his own being the dangerous product of Shakti apart from Shiva. His own being then encompassed and subsumed within itself all of the forces necessary to rule. Yet the capacity for reflexivity was not yet reflexive action. The capacity to discriminate was not yet the act of discrimination—that is, the specification of the functions that operate the detailed workings of the world. These were not objectivated for him, and therefore he did not yet relate to them as the Other that also was integral to the self.

To put this point crudely: in order to become aware that his being was identical to the cosmos, the deity had to incorporate the cosmos within himself. Then, in order to discriminate among the forces and functions that kept the cosmos in operation, he had to spew these forth, to objectivate them, and to recognize that they were all integral to his own being, arranged in accordance with the design of the gods. Then he could act through them in routine ways through the mediation of shakti.

[62] Bryan Pfaffenberger, "The Kataragama Pilgrimage: Hindu-Buddhist Interaction and Its Significance in Sri Lanka's Polyethnic System," *Journal of Asian Studies* 38 (1979): 253–70, and *Caste in Tamil Culture: Religious Foundations of Sudra Domination in Tamil Sri Lanka* (Syracuse, N.Y.: Syracuse University, Maxwell School of Citizenship and Public Affairs, 1982), pp. 114–21. Pfaffenberger does not mention that the Jaffna region is one of intensive Shaivite reformism, begun around the turn of the century, that has succeeded in bringing rituals in numerous temples into line with Agamic procedures. See Perinbanayagam (n. 9 above), pp. 54–60.

He first wed Devayanai, thus creating a relationship with a portion of himself that is extruded, objectivated, and hence known to him in an intersubjective sense. Devayanai's qualities are preeminently those of rulership, fixity, and stability.[63] She is the embodiment of chastity and purity, of *kaṟpu*, and of a formal relationship between the divine and the worshiper. His marriage "cools" the deity. He and his consort dwell in the abode of the gods. He is regally distant from his worshipers. And nothing happens. When the deity and Devayanai are together in the abode of the gods there is a slowing down of cosmic forces and operations. The deity interacts with, and is made over in the image of, Devayanai's rajasik qualities and *kriyā* shakti. This first marriage, this initial devolution of cosmic hierarchy, sets normative parameters on the condition of the phenomenal world. Nonetheless, by itself this dyad would produce entropy and eventually disintegration.

Hence the deity's pursuit of Vaḷḷi. She is his objectification of tamasik qualities and of the force of *icchā* shakti. Vaḷḷi is associated with movement[64] and, one may add, with the speeding up of cosmic forces and operations. And, as he approaches her, he is made over in her image. He again becomes multiform, a shape changer, a trickster, and a destroyer who is deluded and mad with desire. He is furious almost to the point of self-destruction: his threat to ride the hobbyhorse is also one of self-castration and thereby menaces the cosmos with which his being is identified. Yet in this episode, he no longer is as destructive as he was during his evolution, especially during the asura wars. For now there is comparative stability in his encompassment and within the normative boundaries that he has placed on the world. He only threatens self-destruction, and although he kills his pursuers, he later revives them.

But his relationship with Vaḷḷi, as numerous scholars have commented, tests and breaks the conventions established by his first marriage, at least in the sense that he becomes eminently immanent and accessible. Hence, the importance of his proper marriage to Vaḷḷi, for this brings her under his control, as a wife comes beneath that of her husband, a subject under that of his lord, a devotee under that of his god. Nevertheless, the quality of control, as with his first consort, always is relational. When he is close to Vaḷḷi, without mediation, his character takes on attributes of her qualities and forces. Therefore, should he dwell only with her, in this further

---

63 See Beck, "A Praise-Poem for Murugan" (n. 42 above), p. 111.
64 Ibid.

devolution in the hierarchy of the phenomenal world, he would be close to his devotees but incapable of rulership. And once more disintegration would result. The text closes off the degree of devolution of the deity by returning the couple to the abode of the gods, there to live in harmony with Devayanai.

This second marriage marks the completion and integration of the objectification of cosmic forces that act on māyā. The deity discriminates among and controls the opposing forces of *kriyā* and *icchā*: in a sense, he is a divine homeostat governing the slowing down and speeding up of cosmic operations. Therefore he transcends these two forces to epitomize *jñāna* shakti: true perception, illumination, and wisdom that is capable of bestowing grace on the entire cosmos. In this condition, the deity is not personified, for he is an abstract Godhead that approaches the absence of selfhood of the attributeless Brahman. This, too, is dangerous and remains hypothetical, for were the deity to bestow grace on the cosmos and release it from the cycle of karma, the phenomenal world would cease to exist.

The deity's relationship with Vaḷḷi also makes a place for bhakti devotion in this cosmology in a way that is not so nearly self-evident as it may seem. There are, in fact, two great bhaktas in this purana. One, often overlooked, is Surapadman; the other is Vaḷḷi. Surapadman's devotion is part of the radical asymmetry of Shiva and Shakti, of cosmic destruction. He is the product of Shakti uncontrolled, and this force has some affinity to the incrementing of inner heat in the ascetic. Surapadman, of course, pursues Shiva, here the great ascetic. He does so in an ascetic mode, one that bears no small resemblance to that which Hardy calls "intellectual bhakti," or bhakti-yoga.[65] Hardy considers intellectual bhakti to have been a sort of "static service" to the divine that was superseded, at least in Vaishnava Alvar mysticism, by an emotional bhaktism that emphasized the passionate drive to unite in love with the deity.[66] If my interpretation is valid, then the text ironically links the bhakti-yoga of Surapadman to egoism, materialism, and disintegration. By contrast, the devotion of Vaḷḷi is wholly in the spirit of emotional bhakti, of Shakti as the product of the principle of Shiva. It is spiritual, beneficial, and completes the balance of cosmic forces that brings the phenomenal world to fruition. Of course, it is also supportive of the karmic cycle and dharmic order.

---

[65] Hardy (n. 27 above), p. 291.

[66] Ibid., p. 465.

Encompassment and Immanence: Acting in the World

The journey of the *Kandapurāṇam* takes the reader (or listener) through the fragmentation of cosmic order, through the evolution anew of hierarchy and encompassment, and then down through the devolution of the deity and the delineation of the specifics of this cosmic schema. But the text does not go on to tell us how this order works on a mundane basis, through a deity who is simultaneously both encompassing and immanent. Yet there are connotations of these workings, on the one hand, in the logic of the text itself and, on the other, in the perceptions of present-day Tamil worshipers of Murugan. In this section I will point briefly to certain of these implications.

Texts like the *Kandapurāṇam* state that the deity is married to each consort: the tāli is tied for each. As such, the deity fully controls his two shaktis. In this condition he, and the entire triad, are perfectly equilibrated. All knowing, he is not moved to intervene in the mundane doings of the world. This condition is akin to one sort of stasis that is antithetical to the concerns, interests, and desires of the phenomenal world.

The value of this static condition of being is contradicted by a wide array of beliefs, myths, and texts, which in their own ways seek to activate continually the forces embodied by the triad. God must be, at one and the same time, encompassing and encompassed, holistic and specialized, the regal ruler and the involved personal guide of the devotee. As noted, there is no paradox in this since, according to Shaivism, God acts on the māyā of the world (and so on himself) through his more specialized shaktis. Through them he is mediated into the cosmos. Apparently, God does not come, as it were, face-to-face with himself unless his aspects or conditions are in a relation of hierarchy, one to the other, so that one subsumes another.

The stasis of equipose is unlikely because his consorts, more so than he himself in the abstract, are identified strongly with place or location, in a spatial sense that is also hierarchical. The deity and his consorts must live somewhere in the world, whether together or apart. The South Indian conception of the divine rests, in no small measure, on the immanence of deities. In the case of Murugan, his immanence is tied closely to that of his consorts, and theirs to place. As the daughter of Indra, Devayanai's affinity is to the abode of the gods. As the daughter of ascetic and doe, raised by hunters, Vaḷḷi's is to the wilderness. Murugan can live in one or the other, he can move back and forth, or he can go off alone as an ascetic. In the latter instance, there are myths and tales, like that of Parvati's offering the *vel* to him, that draw him back into interaction with the forces of shakti. Therefore, wherever the deity dwells in the world, whether as part of a dyad or triad, his qualities, and so the balance of his being, shift in the direction of that shakti whose natural location becomes his home. Thus, conditioned by spatial location, differing degrees of imbalance are built into relationships within the triad. In turn, this creates a continuum for the shifting balance of forces and qualities that compose the deity.

Consorts are anchored spatially and can move only through the mediation of the deity. He is mobile yet conditioned by the qualities of his consorts. The following combinations of dwelling place and consort are offered by the text and by other myths. First, he lives with Devayanai in the abode of the gods. The emphasis here is on forces of rulership and stability—of *kriyā* shakti and rajasik qualities. There are no startling occurrences, and the deity is on his way to becoming otiose. This is the state of affairs after he weds Devayanai and before he woos Vaḷḷi. Second, the entire triad lives in the abode of the gods. This is the end of the text, at which point the deity embodies the above qualities, leavened by those of Vaḷḷi. Third, the entire triad lives in the wilderness. This is the case after the deity weds Vaḷḷi and, in Sri Lankan tradition, settles with her at Kataragama. After Devayanai tries in vain to persuade him to return to the abode of the gods,[67] she joins the two at Kataragama. The emphasis here is on the force of *icchā* shakti and tamasik qualities, leavened by those of Devayanai. Here the deity is close to his devotees, and they do severe but loving austerities in the spirit of bhakti. Fourth, the deity dwells only with Vaḷḷi in the wilderness. This is expressed in the courtship of Vaḷḷi and is the case in the myths of Kataragama before Devayanai joins them there. The dyad, in this location, is fraught with the action of transformation.

Under no circumstances does the deity dwell only with Vaḷḷi in the abode of the gods, nor only with Devayanai in the wilderness. Nor do the two consorts live together, without the deity. These three combinations are logical possibilities that are realized neither in texts nor, apparently, in oral traditions. This is evidence of the tensile connection between consort and place: each is a function of the other in the phenomenal world.

[67] Such efforts are imbued with transformation and destruction. See Paul Wirz, *Kataragama: The Holiest Place in Ceylon* (Colombo: Lake House Publishers, 1966); and Ponnambalam Arunachalam, "The Worship of Muruka or Skanda (the Kataragam God)," in his *Studies and Transactions, Philosophical and Religious* (Colombo: Colombo Apothecaries, 1937).

The presence of an additional consort, in the location identified with the other, serves to modify the intensity of this relationship.

All of this suggests that a state of perfect equipose in the structure of the triad, and so of the cosmos, is virtually impossible since this necessarily would be conditioned by the metaphysics of place and the premises of shakti (including its absence, should the deity's condition be that of asceticism), as well as by perennial seasonal changes signifying alterations in the quality of the world. But these are more abstract and implicit principles in this tradition of belief. In a text like the *Paripāṭal*, there is open conflict and battle over the deity, between Devayanai's royal warriors and Vaḷḷi's hunters, won by the latter.[68] And, in the folk tradition of *ecal* (mutual insult) literature, Devayanai and Vaḷḷi are fiercely antagonistic toward one another. In my terms, such interactions unbalance the forces and qualities of the triad in mundane and ongoing ways.

There is also some evidence that believers today, through their perceptions of the divine, unbalance the forces of the triad, especially through the subversion of the figure of Vaḷḷi.[69] Worshipers stated that the courtship of the deity and Vaḷḷi was never consummated through marriage since the tāli was not tied. Some obstacle always intervened and the wedding had to be postponed to the following year.[70]

Informants said that, in their previous births, Devayanai was the elder sister and Vaḷḷi the younger. Devayanai, as the elder, had the first right to marry Murugan.[71] But Vaḷḷi tried to seduce Murugan, and Devayanai cursed her to be reborn in the jungle. One version blamed Vaḷḷi more explicitly. Shiva, the eternal brahmacārin, had in his care seven virgins who plucked flowers for him. The youngest of these seduced Murugan. The enraged Shiva turned this girl into a plant creeper (*vaḷḷi*) that was eaten by a doe, who in turn birthed the girl, Vaḷḷi. In this life, continued informants, Vaḷḷi

---

68 Kamil Zvelebil, "Vaḷḷi and Murugan—A Dravidian Myth," *Indo-Iranian Journal* 19 (1977): 227–46, esp. 236.

69 Information on the views of worshipers was collected during and after the Vel Festival in Colombo, Sri Lanka, in 1979, by Bruce Kapferer, E. Balasundaram, Chandra Vitarana, and myself. This occasion is in honor of Murugan and is the most extensive Tamil festival in Colombo.

70 One implication was that Ganesha blocked the wedding. In this there may be a motif of revenge since often the deity is held responsible for Ganesha's elephant head. Once more, raging feelings cluster around the relationship of the deity and Vaḷḷi.

71 Beck, "A Praise-Poem for Murugan," p. 108, comments on the importance for Tamils that an elder sister marry before her younger female sibling. The marriage of the elder establishes the normative value of a correct marriage. Then the younger sister may be given more latitude in hers.

repeats her deed and tries to steal Murugan from Devayanai. Therefore Vaḷḷi continually must suffer, and so she cannot wed Murugan.

Others stated that Vaḷḷi was the daughter of a king. Her birth was not auspicious, and an astrologer predicted that misfortune would befall the royal household were she to remain there. Her father had her placed in a closed box and floated down the river. She was discovered, in various versions, either by fisherfolk, gypsies, or wild Veddahs and raised by them. Shulman writes that temple myths of Tamil Nadu emphasize the violent and creative power of the virgin goddess that should be preserved but also contained, given its danger. "This idea," he adds, "gives us the popular idea of the goddess locked in a box."[72] Thus Vaḷḷi is contained and is prevented from consummating her marriage to the deity. She remains tamasik, the embodiment of *icchā* shakti, uncontrolled by marriage. She is powerful, violent, seductive, and passionate, and she is called the "illegal wife" or the "stolen wife."

Popular perceptions of Vaḷḷi construe her as an improper female who lacks chastity and modesty, and who is eminently desirable. Although she and the deity desire to wed, her fate is to be punished for past sins. And, given her tamasik qualities and force of *icchā* shakti, she cannot rise above her own drives and tendencies. Therefore Vaḷḷi's tāli is not tied by the deity. She remains comparatively autonomous, neither receiving the protection of a husband nor coming fully under his control. Therefore her own power grows (she is considered dangerous), she continues to exert a greater attraction over the deity than does Devayanai, and she pulls him in her direction. Believers hedge this relationship with safeguards. To posit an unmediated dyadic relationship solely between the deity and Vaḷḷi, in her natural location, would be exceedingly precarious. This situation is perhaps equivalent to the deity losing his sense of self within the soul of the devotee, rather than the converse. Believers value the triad that dwells in the abode of the gods, where the deity subsumes the attributes of his consorts. Place and relationship stabilize his composition as comparatively homogeneous and benign, and his concern is more with the regulation of the cosmos. But he does not tie the tāli of Vaḷḷi, and so he is attracted continually to the furious and tender aspect of being that she embodies.

These popular understandings do not leave the activation of the deity to seasonal cycles, to entropy and its regeneration. Instead, these views build

---

72 Shulman, *Tamil Temple Myths* (n. 3 above), p. 140.

into the triad an ongoing moment of momentum of imbalance that shifts the deity in the direction of his devotees, while he also remains stabilized by Devayanai in the abode of the gods. The logic of his composition is one of reconstitution along a continuous hierarchy with immanence at one extreme and transcendence at the other. These remarks require certain qualifications and emendations, which I will now provide.

Conclusion

Let me try to develop further the sort of cosmological hierarchy I think the *Kandapurāṇam* implies. It is a hierarchy without stable anchor, neither at its zenith nor at its nadir. At either of these extremes it would endanger its own phenomenal existence. At its zenith, divine principles involute, retracting from reflexive self-awareness into reflectiveness. Ultimately all principles and components contract, unite and homogenize, losing their differentiation and specialization. The result is a "loss of self" in the sense of an incapacity to distinguish between self and other, although remaining fully aware. This condition of divine contraction is akin to the Indian idea of *nivṛtti*, of the dissolution of an active, particulate cosmos. At the nadir of the cosmic hierarchy, divine principles continue to expand through differentiation and specialization (akin to the Indian idea of *pravṛtti*), are ensnared in the phenomenal reality of a multitude of selves and others, drift into the reflectiveness of passionate attachment, and lose their sense of critical reflexivity. Thus, with further devolution, the divine theoretically could lose its sense of self in the soul of the devotee. The result would be fragmentation and, ultimately, dissolution. Although clearly hierarchical, this setup does have a sense of circularity to it—for zenith and nadir ultimately may meet. One may say that when the dimension of time is added to this hierarchical structure, it comes to resemble the emergence and the retreat of the great cycles of cosmic time in Hinduism—themselves the products of the interplay of *nivṛtti* and *pravṛtti*, of the slowing down and the speeding up of cosmic principles.

The operational hierarchy, the planes and levels of the continuum that signify the workings of the phenomenal world, depend for their coherence on the conscious self-awareness of divine principles. Therefore, the survival of the hierarchy depends on its ongoing self-transformation to avoid either the over-contraction or the over-expansion of the divine at the extremities. The divine must stay within the "reflexive zone" of the cosmic hierarchy, and to do so it can never be at rest.

This hierarchical continuum, one should add, is one of near-infinite gradations whose appearance of stability depends on continuous movement through its ontological categories of greater and lesser encompassment.[73] Thus the iconic and metaphoric expressions of divine principles are at best frozen summations of cosmic processes. Therefore, the oft-reported dualism (or the "splitting") of Hindu deities may be ontologically misleading, albeit iconographically accurate.

There are three major ontological categories or principles of being in the *Kandapurāṇam*: Shiva, Shakti, and asura (or, the "demonic"), in this descending order of encompassment. Kapferer argues that, "given that deities conceived as high in the rank order subsume lower orders in their being, it follows that these deities can manifest themselves at those various levels which they subsume."[74] The highest principle, the most encompassing, reconstitutes itself at all levels of the hierarchy. Shiva creates Shakti, shaktis, and asuras and reconstitutes himself in the obverse direction, at each higher level. But a category constituted at a lower level can reconstitute itself only on still lower levels that it encompasses. Shakti creates asuras and reconstitutes herself (as Māyā) on this lower plane, but Shakti cannot transform herself into Shiva, although Shiva can draw Shakti back into his own being. The asuras, constituted at a still lower level, cannot move up the hierarchy to encompass higher levels without destruction.

Thus the deity can devolve into a refraction of himself (Shakti), into multiple refractions (shaktis), and so on, and he can reconstitute himself holistically by ascending again through these levels. Similarly, Shakti can devolve through her descending aspects—*jñāna, kriyā, icchā*—which then can reconstitute themselves as Shakti. But no principle or category of being can rise above its natural place in the hierarchy, the location where it was constituted or came into being, without disrupting the connection between elements that themselves compose the higher encompassment. Thus the attempt by the asuras to ascend the hierarchy, by usurping the natural location of the gods, turns cosmic order into an incoherent aggregation of disconnected elements[75]—in other words, into ramified disorder. In these terms, the asymmetries in the myth, discussed earlier, are hierarchical imbalances of Shakti claiming at least equality with Shiva, of

---

[73] These comments and others owe their impetus to my understanding of Kapferer's analysis (n. 39 above) of a Sinhala cosmology in Southern Sri Lanka, one that resonates strongly with Southern Hinduism. See pp. 111–28.

[74] Ibid., p. 124.

[75] Ibid., p. 125.

the asuras superseding the gods. Still, the fruitful operation of the cosmos depends on the devolving transformations of higher levels of encompassment, which thereby enables types of beings to ascend once again to their highest place in the cosmic hierarchy.

But a signal irony of this puranic cosmology is that absolute control by the highest level of encompassment produces cosmic death—the cessation of action, of karmic processes, of the opportunities for the soul to attain salvation. Similarly, the utter absence of control, of abandonment, also produces cosmic death through fragmentation at the opposite end of the spectrum. So long as it works, there are no free agents in this cosmology, only complex interdependencies. That is, the cosmology is highly relational: principles and categories, cosmic beings of various sorts, only have significance through their relationships with others of their kind. This is borne out by the connections between Shanmugan ~ Murugan and his two consorts. The shifting balance of these relationships, I argued, keep him within, but moving through, the reflexive zone of hierarchy that is necessary to the fruitful operation of the cosmos. In other words, these relationships and their dynamics are the cosmos; it is their ongoing constitution and reconstitution that gives the cosmos its existence.

Therefore, this cosmos is one that depends strongly on mediation. Earlier I suggested that God apparently does not come face-to-face with himself since his relationships with aspects of himself are mediated by other beings. But hypothetically he can enter into such an unmediated relationship, both at the zenith and at the nadir of the cosmic hierarchy. In both instances—the Brahman absorbing the Atman, the soul absorbing the divine—the result is the disintegration of cosmic order. One might argue that it is the hierarchical organization of the cosmos, and the mediated transformations among its levels, that keep the Godhead from completing itself and so ending the existence of the world.

The requirement of mediated relationships for the maintenance of the cosmic hierarchy also brings to the fore the significance of three sorts of structural arrangements and their relative capacities for mediation. These are the monad, the dyad, and the triad. In South Indian cosmogony the distance between monad and triad is that between passive encompassment and active specificity. The monad, Shiva as the undifferentiated cosmos, is incapable of self-realization or objectivation and so bears only a futile relationship to the phenomenal world. Without mediation, without his shakti—the formation of a dyad—he subsumes all but acts on little. As a dyad, the interplay of Shiva and Shakti is generative but is subject to severe restraints that make their relationship comparatively crude and forthright. This dyad permits three basic kinds of relationship.

One is union, or consensus, between the two. A second is conflict, or opposition, between them. And a third is their separation from one another, a situation of mutual negation. The first is generative, the second cries out for equilibration, and the third is perhaps the most dangerous and destructive. The essential weakness of the monad is that it has little or nothing to contrast itself with and so lacks reflexive self-awareness and the ability to act. The parallel limitation of the dyad is that it lacks the capability to mediate itself. The genius of South Indian cosmogony is its attention to the significance of the triad, an immensely more sophisticated and self-regulating structure. Although the three eyes of Shiva are pan-Indian, it is in South India that the deity married to two consorts triumphs unambiguously over the ascetic divine.

The structural advantages of triad over dyad are clear. The triad permits—indeed, encourages—the formation of coalitions that bring into being different gradients of emphasis of the divine in relation to humankind. As various stories of the deity and his two consorts clarify, these coalitions shift, and, as this happens, the condition of the divine changes accordingly. Moreover, inherent in the structure of the triad is the notion of mediation. The deity mediates between his two consorts, while each of the latter can mediate between her opposite number and the deity. The result is a wider range of subtle and complex permutations that allow the deity to subsume all but that are more compatible with the complicated experiences of humankind. Again, the simultaneity of these relationships is permitted by the hierarchization of the conditions of divinity. These move in fine gradations between the sublime ruler and its counterpart of desire, but they depend on the mediation of the consorts for their existence.

The *Kandapurāṇam* may be thought of as a dialogue, or multilogue, between various points of view, some opposed and some shading into one another. The dialogue explicates the consequences for cosmic order of the alternative relationships among these points of view. Although the text is slanted strongly toward the validity of the position that it espouses, it still brings that position into question and pursues its contradictions until reaching their confluence or eradication. This attribute should be a bench mark for the arousal of reflexivity in myth, as it is in ritual.[76] The text, taken as a whole, seems inherently reflexive in its composition.

---

[76] See Don Handelman, "The Ritual-Clown: Attributes and Affinities," *Anthropos* 76 (1981): 321–70; Don Handelman and Bruce Kapferer, "Symbolic Types, Mediation and the Transformation of Ritual Context: Sinhalese Demons and Tewa Clowns," *Semiotica* 26 (1980): 41–71.

One may do structuralist or phenomenological analyses of any myth in order to unearth its “deep” structure. Yet the deep structure of the *Kandapurāṇam* lies much closer to the surface of the story. I would suggest that this is the case because the complete text demands a more reflexive stance on the part of reader or listener. This is in keeping with an abstract message that pervades the myth: that, in any full sense, only a reflexive cosmos lives; its obverse dies. In turn, this fits well with the Tamil sense of the purpose of existence.

CHAPTER THREE

THE GUISES OF THE GODDESS AND THE TRANSFORMATION OF THE MALE: GANGAMMA'S VISIT TO TIRUPATI, AND THE CONTINUUM OF GENDER[1]

Don Handelman

Telugu Tuesdays are uncertain times. On the fourth Tuesday after the Tamil New Year the Goddess Gangamma visits Tirupati, a small city in the Cittoor District of Andhra Pradesh. Tirupati spreads beneath the ridge on which Venkatesvara (Visnu as the Lord of Venkatam Mountain) has his ancient and wealthy temple. Gangamma is spoken of as his sister (e.g. Cox 1881: 323). The most prominent temple in Tirupati itself is that of Govinda Raja (also Visnu, e.g. Champakalakshmi 1981: 264), the elder brother of Venkatesvara, to whom he has lent money for his dowry. The founding of the Govinda Raja Temple in the twelfth century is often attributed to the sage Ramanuja; the town of Tirupati grew around the Temple precincts (Reddy and Reddy 1990: 145). This area of the modern city, with its narrow *gullies* and caste-named streets, is sometimes referred to as 'Old Tirupati'.

Gangamma's presence swells in Tirupati during the month of Vaisakha (May–June), a period of dry, searing heat, wind and dust, in the past of raging epidemics, yet also the moment when transplanted rice shoots in the paddy fields are about to sprout in anticipation of the monsoon rains. Her visit is explicitly related to human fertility, the fertility of the soil, bringing rain, protection against disease and general well-being. Gangamma moves in with a family of weavers (*Kaikala*) whom she has enjoined to enact the story of her sojourn in Tirupati. Her myth is outlined in this essay.

During the next week these weavers, with the aid of other castes, embody the story of the Goddess's chase, destruction and transformation of the male. Just as the Goddess concealed herself in a sequence of guises

[1] My thanks to the Tirupati collective—A. Anand, Joyce Flueckiger, V. Narayana Rao and David Shulman—for their encouragement, warmth and spirit; and to David Shulman for his sustenance and for a close reading of this essay. Living with Lea Shamgar-Handelman has awakened me over the years to the manifold weaves and textures of gender.

to seek out the lusting lord who sexually despoiled all virgins, all 'flowering blossoms' (*pushpavatis*), so too the weavers take on these guises of the Goddess in her search for vengeance. Embodied in her guises the Goddess comes into increasingly close contact with human beings. Gangamma's quest for the lascivious lord brings her into direct, unmediated contact with numerous households in Old Tirupati, and especially with their women. The entire cycle of guises is primarily a form of domestic ritual that brings the Goddess and families increasingly closer to each other.

In terms that are not native to Tirupati, the progression of guises constitutes a causal sequence of ritual action (Handelman 1990: 22–62) that makes Gangamma increasingly present and active. Moreover, the Goddess's relentless pursuit of the male corresponds with her emergence into full self-awareness in the world of human beings. As the male is destroyed the cosmos of the Jatra is feminized, and this female cosmos is one of bounty for all.[2] The guises of the Goddess generate the transformation of the male. The Gangamma Jatra raises contradictions and ambiguities between the metaphysics and social norms of south Indian gender—between gender conceived as a continuum and its abrupt rupture into sharp difference in accordance with cultural roles. I will address these issues in the concluding section in which I argue that the vision of gender in the Gangamma Jatra contests those usually offered for south Indian women, goddesses and others, but that such alternatives (in this instance imagined through ritual) are not available in social life. I contend further that processes of gender and cosmos in this Jatra may have particular relevance for castes of the left-hand. In the concluding section my interpretation depends on taking the Goddess's point of view, with all the hubris this entails. In this regard, Margaret Egnor's (1984: 24) comment is to the point: 'Hindu deities are not only symbols or tools which are manipulated by human beings to express certain ideas—such deities are actors, with a will of their own.' Despite the supposed florescence in anthropology of the native's point of view, south Indian goddesses in particular have had a poor, overly selective representation from their interlocutors.

Gangamma's visit to Tirupati ignites other domains of ritual activity at temples, within homes, and through the practice of personal vows to the Goddess. At the close of the Jatra the sequence of guises intersects

[2] The commonest English translation of Jatra is 'festival'. This translation is misguided and misleading. The signal connotation of Jatra is 'movement', and this is crucial to the rites of Gangamma. Therefore I use the term Jatra throughout this work.

powerfully with rituals at temples. Therefore the general rhythms of temple-related rituals are noted briefly in the body of this essay.

Gangamma in Tirupati

Gangamma has two major temples in Tirupati (a third that comes into prominence during the Jatra is discussed later). The character of the Goddess of each shrine and the linkage between them delineate the patterning of Gangamma's relationship to Tirupati and to her worshippers. These goddesses are Gangamma of Tallapaka (a town some distance to the southwest of the city, from where, according to one myth, the Goddess was brought to Tirupati by the Telugu poet-sage, Annamacharya), and Gangamma of Tatayyagunta (a periphery of the city). For the moment I discuss them in conjunction, but shortly each will be highlighted separately. Within each shrine the true form of the goddess is a dark, stone head. Each goddess has her stone feet pointing outward, in the courtyard outside the temple. The feet of each pair meet at heel and toe, with an open, oval slit mid-length between them. During the Jatra each austere stone head with her darkly etched features is given a face of lighter, brownish clay carefully made up, her fangs hidden beneath vermilion lips that verge on a satisfied smile. Tallapaka and Tatayyagunta are described respectively as the older and younger sister. Tallapaka has her own children, Tatayyagunta is barren. Pressed by Tatayyagunta for children, Tallapaka hid hers under a winnowing basket, but eventually relinquished half her offspring to her younger sister. The stone children of each surround them in their respective shrines.

The modest temple of Tallapaka—a tiny one-room shrine forefronted by a narrow courtyard—is located in the very middle of a busy thoroughfare next to a major intersection of three bustling city streets. Tallapaka is a 'middle-of-the-street' Gangamma. She has her own children and does not bring disease. Her 'middle' status is significant. Streets, and especially intersections in south India, are perceived as loci of movement between cosmic planes, and therefore require special guardianship against undesired intrusions. The 'middle' also connotes a cosmic axis (e.g. Mount Meru), an earth navel (e.g. the Śaiva lingam) or the *garbhagṛha*, the 'womb house' or innermost sanctum of a deity within its temple home (Lannoy 1971: 40) or, one may well add, within itself. The innermost depth is also the most transcendent of cosmic levels, so that 'deep within' (indeed, deep within one's self) and 'up high' are homologous. David Shulman (personal communication) comments that the 'centre' in Hinduism is a locus of movement, ambiguity and transformation, the place where doubts about reality arise.

At the same time, Tallapaka's 'middle' qualities also connote the innerness of domestic space and family intimacy. Domestic rituals within the household to this Goddess call her 'Middle-of-the-House Gangamma'; while the last public rite for Gangamma during her annual, solar cycle is in the month of Karttik, in mid-December at the end of harvest, when every street in Old Tirupati should ideally make and worship a clay Middle-of-the-Street Gangamma. Tallapaka's middle-of-the-street epistemology fuses axiality and guardianship with innerness and the domestic domain.[3]

Tatayyagunta lives in the largest goddess temple in Tirupati, with a courtyard precinct of corresponding size. She is on the periphery of the city (Reddy 1976), brings epidemic disease as well as protects people against diseases. In location and ethos, she may appear closer to numerous other south Indian goddesses who are angry, violent mothers of disease, and both protectors and violators of territorial boundaries (Beck 1981; Bradford 1983; Good 1983: 239; Moffatt 1979: 252–81; Reiniche 1987; Van Der Hoek 1979). Yet Tatayyagunta does not seem to inspire emotions of dread—for example the deep fear and euphemistic use of 'mother'—that violent south Indian goddesses are said to evoke.[4]

Descriptions of angry Indian goddesses attribute their violence to the absence of a consort (e.g. Babb 1970). Many of these goddesses are ferocious conqueresses until tamed, married, and made docile and passive in relation to their husbands (e.g. Fuller 1980; Shulman 1980; Ramanujan 1986). By contrast, the marital status of Gangamma (in whichever of her forms) is not an issue of any import. The Gangammas wear *talis*, but there is no ritual act in the Jatra that explicates or implicates their marriage (in contrast, see Beck 1981; Moreno and Marriott 1989: 158–64).[5] Undoubtedly the Gangammas are potentially dangerous in their active, hungry presence. Yet they (especially Tallapaka) are intimately and even prosaically domestic beings. Thus their wearing the tali may be linked first and foremost to their domesticity, rather than to the issue of marital status. As a

---

[3] See also Elmore's (1984: 40) brief reference to a 'middle street sakti' (*nadividhi śakti*) who 'is not quite so terrible as some of the other [village] saktis'.

[4] Generally, south Indian Goddesses who protect territory and who bring disease are distant in conception from north Indian disease Goddesses (e.g. Wadley 1980; Kolenda 1981; and to a degree, Hershman 1977).

[5] The tali is the wedding badge tied by the groom around the neck of the bride, signifying his superordination over her and the extension of his protection to her.

2. Tatayyagunta Gangamma Adorned (Joyce Flueckiger).

woman worshipper of Tallapaka said: 'It's seemly for a woman with children to wear a tali.' And if anything, 'husbands' become increasingly conspicuous by their absence during the progression of the Jatra. Especially for women, Gangamma has something of the status of a good and valued friend, loved and admired (Egnor 1984: 28).

The location and ethos of these Gangamma temples show the same goddess in the middle or centre and at the periphery. The differences between the two are of degree, of locations on a continuum of female being and power with perhaps a certain division of labour amongst them. The same pattern is reproduced within the Gangamma Jatra. This pattern also highlights that whatever her form or aspect, Gangamma is a powerful, independent female in her own right. Without a consort, she is 'very much the mistress of her own feelings and actions' (Hanchett 1988: 187). Through its sequential progression her Jatra increasingly brings into focus a cosmos that exfoliates her being.

Gangamma's Story

The narrative that empowers Gangamma's annual visit to Tirupati tells of her as a ferocious female power (*śakti*) driven by hunger, swallowing people who answer her knock in the middle of the night. Tamed by Ramanuja, she turned into a ten-year-old whom he took to Tirupati and left on a streetcorner. Later a pavilion (*maṇḍapam*) announcing her entry was built on that spot, and she lives there in the form of an oval stone called Catu (announcement) Gangamma. Gangamma also looked for a home with shepherds (Gollas) and fishermen (Palli), but could not stand the smell of their food. She decided that only the weavers could bear her, and one night she accosted the great-grandfather of today's weavers. He refused out of modesty to admit her—a young, beautiful, unattached virgin—into his home. Not to be denied, she formally announced her intention in court and moved in.

At that time a Palegadu, a local warrior lord and landowner, was forcing every virgin to sleep with him the night before her wedding. He terrified everyone. Spying Gangamma drying her hair, he lusted for her. She decided to teach him the lesson of his life. But first she needed to discover where he lived. Each day she and her sister took another guise and went through Tirupati searching for him. The day before Gangamma killed the Palegadu she took the guise of an untouchable sweeper (Toti) and went from house to house assuring everyone that they would not be harmed. On the day of the killing she took the guise of a foreign ruler, a 'prince from another place' (Dora), and went to the Palegadu's palace, where he was playing dice with friends. He came forth to fight the Dora, and Gangamma beheaded him. She revived him for a few moments to show him her true form (*viśvarūpa*), telling him: 'This is what a woman really is.' He pleaded for his life, vowing his devotion to her. Gangamma refused, saying he would be just the way he was before, and killed him again.[6] After this she pierced her tongue with a needle to demonstrate her power. Gangamma instructed the weavers how to re-enact her story, the guises to take and how to dress and embody them so that she would return to Tirupati. Then she turned herself into a stone at the Tallapaka shrine.

On its surface the story of Gangamma is a charter for the embodiment of her guises, and gives some explanation to three of these (sweeper, foreign prince and viśvarūpa). The story conveys progression and transformation as the Goddess drops the guise of the prince and reveals herself as the totality of female power, utterly eliminating her male enemy. But the story is concerned solely with Gangamma's origins and her connection with Tirupati, ending with the emergence of the fullness of her being in the world of people. The story refers neither to the additional characters that appear in this ritual sequence following the viśvarūpa, nor to other actions during the Jatra by and for the Goddess. These are crucial to the performance of this sequence and to the well-being of Old Tirupati. In this regard the ritual specialists provide little exegesis, stating that they follow the instructions of Gangamma. Nonetheless the practice of this sequence swells with causal, transformative moments that are profound in their metaphysical consequences.[7]

Gangamma Comes Home

On that fourth Tuesday the weavers bring home Gangamma from the Tallapaka Gangamma Temple. Accompanying her is the red-painted wooden head of the mustachioed Palegadu. Four families of weavers, living close to one another, have hereditary rights in the Jatra. According to

[6] During the Jatra, drummer bards chant the myth of Ādi-Śakti, the supreme female power who creates the cosmos. Feeling lust, she creates sons (Brahma, Visnu) in order to have sex with them. Morally outraged, they refuse and she destroys them. In shaping Siva she meets her match, as he tricks her powers (including the third eye) from her, thereby taming her. There are various versions of this myth in the area of Tirupati and elsewhere (e.g. see Oppert 1978: 472–4, for a version in which Visnu is triumphant).

[7] On the significance of sequential ordering in ritual, see Handelman (1990: 22–62); and on guising and self-identity, Handelman (1984, 1990: 138–59).

the account of their origin they once were a single family (in their great-grandfather's time, when he met Gangamma) that split into the families of two brothers, and then four. Each family hosts the Goddess and enacts her guises during two consecutive Jatras, and then passes this right to the next family. The keys to the main gate of the Govinda Raja Temple also pass with the Jatra from one host family to the next. Thus during a given two-year period, one weaver family hosts Gangamma, its men enact her guises, and are responsible for the opening and closing of the Govinda Raja Temple. The weavers, one may say, embody their caste namesake in shaping the texture of Gangamma's presence in Tirupati, the weave that interlaces the Goddess and her devotees.

The portable, wooden image of Gangamma that enters the weaver home is some two feet in height, red-coloured, angular and fierce, leaning to the left and holding a lemon-tipped sword in her right hand. Her white corneas and black irises are eye-catching, as are her two white fangs. Her shrine is established next to one wall of the dwelling, where she is joined by the permanent household image, a silver-coloured, fanged, symmetrically proportioned head of Gangamma. Soon after, the steel swords of the foreign prince and his minister (*mantrin*) that will be used to kill the Palegadu are placed on either side of the visiting Gangamma. These goddesses, like all the Tirupati deities discussed in this essay, face the east, the auspicious, beneficial direction of the rising sun (Beck 1976: 215). During Gangamma's stay, members of the household observe certain restrictions, refraining from sex, not pounding spices, eating calming foods and going barefoot outside—and of course, inside—the home. Gangamma eats the food of the household. She is not calmed, as an inert object would be. Rather, she takes in the substances offered her, thereby soothing herself. She is perceived always as an autonomous, causal agent.

The Goddess who moves into the centre of the weaver home is Middle-of-the-Street Gangamma. Though the Goddess is in her Tallapaka shrine, from the perspective of the participants, the force of her power shifts into the weaver household. Gangamma herself has returned home. Worshipped and respected, she is seen as a loved member of the family, not someone to fear. In shifting from the temple to the home, she goes directly into intimate, domestic space, with its connotations of fertility and growth. As she moves into the midst of human beings, so begins the recall of how she previously became intimate with them—finding shelter in Tirupati, protecting women against the depredations of the Palegadu.

Wednesday morning, weavers and washermen (Cakali) accompanied by drummers (Pombala) make a *mandala* to enclose space that is referred to as Old Tirupati. Their ritual acts are sometimes described as 'fixing the circle' (*cakrabandhanam*). The enclosed space of the mandala is called an *ūru*, a village. But the Telugu term has connotations of the Tamil term *ūr* (e.g. Daniels 1984: 61–79)—of ancestral soil or ancestral place, with which persons born there have a special relationship of mutual affinity. The sequence of guises dominated by the weavers is enacted within this enclosed space. They say that during the Jatra the Goddess does not like her children, the people of the ūru, to leave this space. From this perspective Gangamma is indeed the goddess of the Tirupati ūru. But the mandala has an additional purpose—to ensure that during the Jatra, Gangamma does not stray from the ūru. She and her children should stay within this ancestral land where they are woven into one another through affinity, and through acts of reciprocity and mutual benefice.

The implications of this formulation penetrate what will be done through the ritual sequence of guises. Gangamma gave the weavers instructions on how to make her present in the ūru by taking her guises. Her intention, her desire, is to be in the ūru with her children. But then, once in the weaver home, why should she leave, why wander away out of the ūru? Thus if she should depart, this would not be intentional. In question therefore is whether at the outset of the Jatra she is fully self-aware of being present in the world of human beings. The implicit agenda of the sequence of guises is to transform Gangamma's sense of self for the benefit of her worshippers, enticing her out of her own centre, her 'middle', into her full presence among human beings. This is also her devolvement from cosmic encompassment to the earthly plane. Though the mandala fixes her, as it were, in space, this by itself neither makes her present nor prevents her from disappearing deep into her self, again distant from the human plane. She must be persuaded continuously to emerge from within herself, seduced into her self so that she remains among people during the Jatra.[8]

During the Jatra, numerous rites are performed at the Gangamma temples of Tallapaka and Tatayyagunta, especially at the latter. In outline, the rhythms of these rites parallel those of the sequence of guising enacted

[8] There are implications here for daily temple rituals that wake, wash, dress and feed the deity. Rather than merely preparing the deity as royalty, the more profound metaphysical connotations of these rites are of persuading the deity to come out of herself or himself, nearing the human plane. Thus the act of closing this enticement into embodiment by showing the deity to herself or himself in a mirror suggests the conscious self-recognition by the deity of cosmic encompassment and therefore the desire to act deliberately in this world.

by the weavers. However these shrines remain apart from this sequence. Nonetheless, within mandalic space the same relationship between a Gangamma of the middle and one of the periphery is reproduced. During the Jatra interaction between these goddesses of mandalic middle and periphery is much more intense than is interaction between Tallapaka and Tatayyagunta during the everyday.

The middle Gangamma of mandalic space is the visiting Tallapaka who has moved into the weaver home. The Gangamma of the periphery is Veshalamma, the Mother of Guises (*vesham*). Her temple is a kilometer's distance west of the weaver home. Veshalamma is also, one may say, the mother of borders, since guise itself is the border of perception between exterior and interior, and the movement between them. Veshalamma too is Gangamma. The only image in her shrine is full-bodied. Yet her true form is also a head, lost when thrown down a deep well by a crazed eunuch (e.g. a male feminized in his loss but not made female, and therefore in disjunction, rather than in harmony, with the female). Most of the guises begin their circuit at Veshalamma, moving inward from the edge, into the ūru, seeming on the surface to invert the progression of guises that leads towards the revelation of the Goddess from beneath the guise, from within. Yet there is no contradiction here. In spatial terms, the movement of the circuit, from the outside inwards, is one that comes closer to the inner sociality of human beings, their domesticity and intimacy. In existential terms, the movement of the Goddess from within her being outwards, comes closer to others and penetrates their being. Thus the spatial and the existential complement one another. The movement inward in space is outward in self, then moving inward again towards others. The transformation within the Goddess through the progression of guises, her emergence from within herself, is mapped and reproduced in space through the movement of guises from Veshalamma into the ūru. I discuss this in the concluding section.

The Goddess Disguised

The circuits of the guises begin on Wednesday morning following the making of the mandala, and continue into daybreak of the following Wednesday. This period divides into two segments of guising, that of the Goddess disguised (Wednesday to Friday) and that of the Goddess revealed (Sunday to Wednesday), linked by an interstitial guise (Saturday) that turns disguise into revelation. In temporal terms, this transformation is done in the middle of the Jatra. The division into two segments corresponds to Tirupati's understanding regarding the presence of the Goddess. It also enables attending to subtle shirts in the Goddess's progressive perception of self and the intensification of emotions surrounding her (the sequence of guises is summarized in Table 1).

The first guise on Wednesday morning is that of a single Bairagi, a male Śaiva ascetic, enacted by a washerman. The Bairagi announces the Jatra in major locations within Old Tirupati. From then on through Friday, each guise appears as a married pair—the female enacted by an unmarried weaver male, and the male enacted by an unmarried washerman. Each pair is understood as the Goddess (disguised as the female of the pair) and her sister (disguised as the male of the pair) going from dwelling to dwelling in search of the Palegadu. The Goddess (or Goddesses) 'comes on' (possesses) those who embody her. The Goddess is, one may say, within the weaver and washerman who wear the disguise she takes on.[9] On Saturday both partners are male—the foreign prince and his minister—enacted now by a weaver and washerman who are both married. This pair behead the Palegadu at daybreak on Sunday. It is in the moment of the killing that the Goddess reveals her unitary self, the shells of her disguises tossed aside. Though these goddesses are a pair, from the outset, it is the disguises as women (embodied by weavers) that are closer to the female core of the Goddess, and that become progressively more central. Parallel to this the male disguises (embodied by washermen) are discarded. The weavers then become the sole embodiments of the Goddess.

All guises are attired in the weaver home. On Thursday and Friday, one pair leaves at daybreak and another in the afternoon. The males of the pair, the washermen, are dressed simply in headcloth, shoulder cloth, dhoti or pantaloons, carrying artefacts (baskets, scales, necklace of vegetables, axe, cowbells) that identify the caste or tribal group of the guise. Stripes or patches of colour and other decorations are applied to their faces and upper bodies. Special care is given to the details of attiring the woman, the Weaver of the pair, to her sari, upper cloth, necklet, necklaces, bangles, anklets and flowers. Her hair is covered by a black cloth covered in turn with gold-coloured ornaments that also hide her ears. Kohl is applied delicately to her eyes and eyelids, beauty spots to her cheeks, and

[9] The washermen enacting the male guises probably have an important purificatory function at the interface between the Goddess and her worshippers. It should be noted that none of the ritualists who perform in the sequence of guises are full-time specialists.

a vermilion stripe to her forehead. Some item of her apparel or accessories (for example, ankle bells) are put for some time before Gangamma, purified over a brazier of incense and, in a transference of identity and power, added to the attire. Attention is given to the calming of the female guise: turmeric is rubbed into her cheeks, giving her skin a yellowish cast, and jasmine and other flowers are twined intricately around her head ornaments and down her back. Before departing, fire from Gangamma's altar is waved (*ārati*) before the couple, and the woman covers her head with a white cloth.

The two walk briskly to the Veshalamma Temple, ten minutes or so in a westerly direction, accompanied by one or more drummers, their drums silent. The full embodiment of most guises requires the joining of Gangamma as a middle goddess (Tallapaka in the weaver home, her embodiment in the guiser) and Gangamma of the periphery. As noted, the latter is Veshalamma, the border between the outside and the inside of the guise. At Veshalamma, the pair greet the Goddess, the woman removes her headcovering, the drummers sound their instruments, and the couple returns at a speedy pace in an easterly direction, moving into the ritual space of the ūru, and focusing attention on the interior of the guise. The pair stop on the pavement outside particular dwellings where family members, mainly women and girls, wait to worship them. The basic form of worship is to wash the couple's feet and to lap the water towards one's face; or to touch the feet and then one's lips. Others also decorate the feet with turmeric and vermilion, and offer bananas, coconuts and the fire of camphor flames. Occasionally the head of a small child is touched to the couple's feet.

On Wednesday afternoon the first couple—snake-charmers—do their circuit through Old Tirupati. They are followed on Thursday morning by shepherds (Gollas, e.g. Sontheimer 1989: 102, 105) and in the afternoon by ruffians (Banda and Bandati). On Friday morning the pair are merchants (Chettis) and in the afternoon, untouchable sweepers (Totis, e.g. scavengers). The circuits of these first guises take some five to seven hours each. Apart from the devout acts of worship by people awaiting them at set stops outside homes, the sight of these disguised Goddesses walking the narrow streets accompanied by drummers does not arouse any particular curiosity or enthusiasm.

The ruffian disguise intimates that the presence of the self of the Goddess in these early guises is still limited. The male ruffian's apparel stresses his masculinity. He wears red and gold striped pants with a white cloth wrapped intricately between buttocks and genitals to shape a codpiece-like

3. Veshalamma (Don Handelman).

protrusion. The weavers say that the woman, Gangamma in the ruffian guise, took a fisherman lover. The lover and husband fought over her, the lover cutting off the husband's left hand and the husband slicing off his wife's nose to disfigure her beauty. Near the Street of the Fishermen (close to Veshalamma Temple) a winnowing basket is attached to the ruffian's arm, simulating a covered stump, and a strip of plaster stuck lengthwise on his wife's nose. In this disguise Gangamma suffers severe facial damage. Given the care with which she always dresses, the Goddess is very sensitive to her own appearance. Her beauty is integral to her sense of completion. In this altercation violence is done to her, she is made ugly by a male, and this undoubtedly injures her sense of self. Nonetheless she does not react to this insult (one that likely equals in magnitude the lascivious pursuit of the Palegadu), behaving as a passive, docile wife punished for her sexual misdeed. Where then is the Goddess in these early guises? Not only is she disguised, but her embodied presence is still distant.

Following on the ruffianess, the apparel of the merchant's wife indicates the need to calm the slowly swelling emergence of the Goddess. She is the first to wear a garland of neem leaves (margosa, a powerful calming agent) and jasmine (also calming). But the general disinterest in the disguised Goddess as she walks the streets changes dramatically with the appearance of the sweepers (Totis). The male sweeper carries a naked,

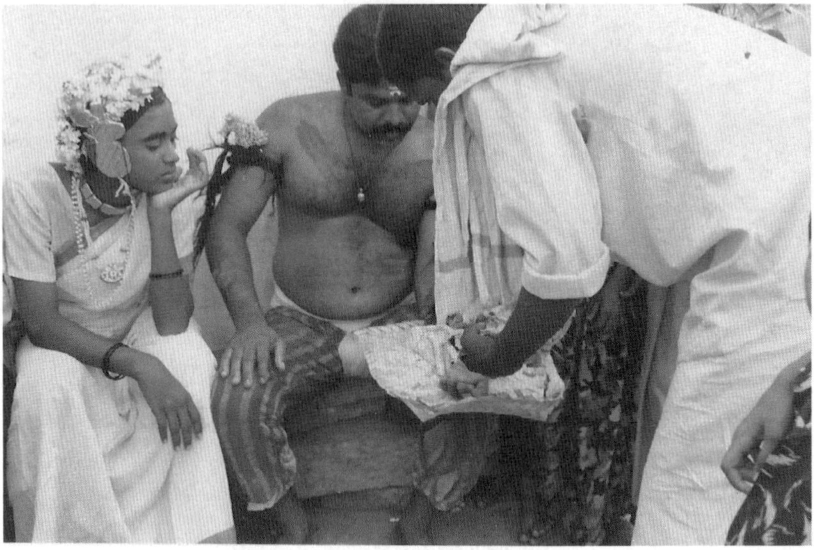

4. Ruffian Guise, Female Ruffian beside Him (Don Handelman).

5. Gollas Arriving at Veshalamma (Don Handelman).

raised sword tipped with a calming lemon, the first weapon to appear among the guises and probably indicative of the Goddess's increasing self-awareness of her pursuit and purpose. The female sweeper's face and body are blackened, and she too has a garland of neem and jasmine around her neck. Following the sweeper, all guises wear this garland. A winnowing basket with five points of vermilion (signifying the five cosmic directions) is placed before Gangamma and purified with incense. The winnowing basket is a gift to Gangamma from her brother, Venkatesvara.

At the temple of Veshalamma (who wears the same garland as the female sweeper) the cosmic power of the winnowing basket is highlighted. The basket is worshipped separately—a camphor flame placed before it, a coconut broken—and then taken up by the female sweeper. Of the disguises so far, the sweeper has the most intense relationship with Veshalamma. According to Gangamma's myth, it was in the sweeper's disguise that she went from house to house assuring people they would not be harmed when she killed the Palegadu. Along the circuit of house stops, people ask the female to clap them on the head with the winnowing basket. For that matter everyone in the street is clamouring for a head tap. An ambulance stops, its crew getting out for claps. A bus stops, its passengers leaning out of windows for taps. The female sweeper is surrounded by small, shifting crowds. People are smiling, grinning, having fun, in a good mood, as they rush about her, tugging at her arms to attract attention. At the mouth of an alley the crowd hems her in, shoving her back, young and old, women carrying infants, children and teenagers, not giving way until they receive her taps. The female sweeper is tall, a black head and arm above the crowd, clapping, tapping. Breaking out of the alley she puts her shoulder into a teenager, loping to escape the crowds and get on with the stops outside homes. During all this the male sweeper is ignored. He waits by himself, holding his phallic weapon upright. The focus is wholly on the female and the blows of her basket.

The Goddess blesses with her winnowing basket.[10] Yet she has been around for the past three days without attracting great attention. Now

---

[10] In the rites and myths of Gangamma, the motif of the winnowing basket has a tacit significance. In the hands of a man, the basket conceals damage to the masculine (e.g. the ruffian twists the winnowing basket into concealing absence and loss, e.g. the stump of an arm). In the hands of a woman, the basket reveals the holism of female power. Tallapaka Gangamma protected her children from her sister under a winnowing basket. The female sweeper who mixes everyone's leavings together probably uses the basket to dispense blessings of fertility. Its shape may also be likened to a vagina, or perhaps a womb. The winnowing basket is the gift of Venkatesvara who in popular myth was a female turned

6. Female Sweeper Swatting Heads (Don Handelman).

there is a vibrant desire for interaction with Gangamma. The people recognize a change in her. She is no longer the wife of the snake-charmer, the shepherd, the merchant. She is no more the meek woman whose nose is cut off by her ruffian husband. As a sweeper, a scavenger of others' leftovers, she is very close to the realities of the human world. She is becoming aware of herself as an independent force with something to give that people really want. People are entering into an emotional, reciprocal relationship with the female, virtually disregarding the male presence of her companion. The female sweeper is the first disguise to give the extraordinary benefice of the Goddess to anyone who requests it.[11] Moreover, apart from those awaiting her at set stops, the numerous beneficiaries along the way are not receiving these claps of the basket in return for acts of formal worship. The Goddess's benefice is available at this moment to anyone

into a male by Ramanuja. Thus the winnowing basket may also signify the movement of female power from one Goddess to another, and so the protection of its holism. It is worth mentioning here that the earliest poetic references to Venkateśvara describe him as half Visnu, half Śiva. The Goddesses of Tirupati have Śaiva connections that cannot be gone into here.

[11] Previous guises distribute benefice at pre-set stops in keeping with the characters of these guises. Thus the shepherdess pours milk, and the female merchant (the male sells vegetables) distributes rice.

who recognizes her within the disguise. The Untouchable Sweeper at the bottom of the caste hierarchy is no respector of categorical, hierarchical differences—she tosses everyone's leavings together. So too is the benefice of the Goddess ultimately intended for everyone. Gangamma is now very close to the surface of her disguise and to the world of human beings. In the progression of guises the interaction between sweepers and people establishes the reality of cosmic power as female, not male.

There is no guise on Saturday morning. Saturday is a time of transformation in the Jatra, the temporal mid-point of the sequence of guising, as its momentum accelerates from concealment to revelation. Three days of the disguised Goddess precede this day of the killing of the male, and three days of the revealed Goddess will follow. By Saturday the Goddess is through disguising herself as a female. She is just that.

From now until the end of the Jatra the Goddess is embodied only in married males.[12] On Saturday afternoon two males appear: the foreign prince and his minister. Within the cosmos created by these rituals, these two figures are the apex of male power. Within them the Goddess has emerged fully. Thus, though the surface of this guise is entirely male, its reality is entirely female. Unlike the previous guises who wander with their uncertain identities, this pair sets out with the vividly declared purpose of beheading the Palegadu, the male. They will be on their way through the streets for some nineteen hours—a further intensification of the presence of the Goddess—killing the Palegadu at daybreak on Sunday morning. The Palegadu will see only the male surface of the guise, and not the female reality beneath.

The prince and his minister are dressed in the colours of Visnu, their clothing white with red sashes at the waist, the prince with a blue sash across the right shoulder. Each has the Vaisnava trident painted in white and red on his forehead. There is a feminine cast to the prince's make-up,

---

[12] In the myth of Gangamma, only the weavers could bear her power. As her presence in Tirupati swells, only married weavers can bear her. There is here some intimation of control through marriage. From the Goddess's point of view, during the initial three days of the Jatra, her first embodiment is a male ascetic distant from marriage, followed by two females (Goddesses) embodying two unmarried males enacting a married couple, the female controlled by the male. During this period, marriage is initially absent and is then entirely on the surface of the guise and the Goddess most distant from this. On Saturday the two Goddesses embody two married males enacting two other males. The subject of marriage is now beneath the surface of the guise, closer to the Goddess. During the last three days the Goddess is wholly on the surface, embodied by married males. In one sense, these men are within her; in another she is within them. To some degree her self has moved into closer conjunction with a condition of marriage in the human world.

especially in his eyes lengthened by kohl. Their steel swords were on either side of Gangamma since she came home, and their hats were also before her for a while. They have the largest entourage of any disguise, three drummers, two standard bearers and another to hold the Palegadu's wooden head that had accompanied Gangamma to the weaver's home.

The drums accompanying previous guises were sounded only on leaving Veshalamma. Now the drums are beaten as soon as the pair leaves the weavers' home. Their swords upright in their fists, the pair set out at a martial pace in the direction of Veshalamma, but do not visit her. Instead, they stop in the street at a point directly in line with her temple. At this boundary the pair are worshipped apart from Veshalamma. They become the analogue of Veshalamma, keying the border between the interior and exterior of ritual space and of the inside and outside of disguise. They consciously control these boundaries that are now open to them. This is a clear index that Gangamma is actively present, just within the surface of the disguise. The Palegadu's palace lies beyond, down the same street. For the present, the pair turn about and head into the ūru to begin their circuit of stops outside homes.

Unlike the tumultuous reception accorded the sweeper the day before, there are few people on the street this Saturday. Nonetheless the stops are lengthy and movement slow. At each stop the Palegadu's head is held aloft while the minister recites his wicked ways, declaring that the prince will behead the Palegadu and stuff his head into his belly. The prince and his minister are greeted royally. Often chairs are put out for them and they are worshipped while seated. Scarves are placed on their shoulders and garlands around their necks. Their feet are washed and touched, and turmeric and vermilion applied. Coconuts are broken before them, and they are offered camphor flames, buttermilk, soft drinks, bananas, coconuts and money. In return the two cut garlands into segments with their swords, and return these to householders. Although they are received outside homes, at times there is an intimacy between these visitors and their hosts, as they chat and banter with one another.

The prince and his minister travel throughout the night. By daybreak they near Veshalamma, but continue straight on past the temple for several hundred metres and around a bend in the street.[13] The Palegadu's wooden head, garlanded with jasmine, is held high. Keeping the head to

[13] The prince and minister also march beyond the ritual periphery, to the homes of families who once lived in Old Tirupati.

7. Dora and Minister Received Outside Home (Joyce Flueckiger).

8. Prince and Minister with Head of Palegadu (Don Handelman).

their right, the prince and his minister circumambulate the head three times and then touch it with their raised swords, accompanied by heightened drumming and ululation. A washerman quickly hides the severed head inside a cloth and departs on the run to Tallapaka Temple, where the head reappears later in the day next to Tallapaka Gangamma. There are no spectators to the killing, apart from the entourage and some youngsters. The important thing is that the ritual is correctly performed.

In the myth of Gangamma, the Palegadu is playing dice when she comes for him.[14] He sees two foreign lords, warriors, and comes out to fight. His perception occluded by his extreme maleness, he fails to see beneath the surface of the male and dies at the hands of the female, who then shows him 'what a woman really is'. The Palegadu embodies not only maleness, but maleness at its extremity of uncontrolled desire that destroys the power of female virginity. He is a threat to order, especially

[14] The role of dice and analogous games in South Asian myth and epic has received more than a little scholarly attention. Much of this is incorrect in its insistence on the homology of the game and the cosmos, such that the player outside and inside the game is considered to have the same level of perception. It is more likely that entering the game lowers consciousness, occludes perception and restricts knowledge. David Shulman and I have written at length on this subject.

that of the domestic domain. To eliminate this maleness, the Goddess raises the status of her disguise to that of the prince, analogous in hierarchy to the Palegadu. The pretense of maleness destroys the male, revealing the cosmic power of the female. In terms of this ritual cosmos more generally, maleness as a total identity is a disguise for a more complex self combining male and female attributes.

Although there is a demonic cast to the Palegadu's unbridled lust, he is an ordinary man. He is neither a demon whom the Goddess marries and kills, nor is he a deity who marries the Goddess and controls her. He is a male unable to perceive the reality of the female. The death of the Palegadu, Gangamma's refusal to let him live as her devotee, turns the ritual-cosmos female and shows the epistemological superficiality of the male who is only that. The surface appearance of maleness (the prince and his minister) cancels maleness (the Palegadu) that perceives only the surface, and the Goddess emerges wholly from within her male guise, consciously and deliberately showing her true form as the dominant principle of this ritual cosmos. No longer are there males of any significance. The true self of the female has risen through maleness, incorporating this into herself. Meanwhile, as the prince is killing the Palegadu, the true form of the Goddess, the Matangi, is visiting Veshalamma.

Gangamma's Feet

On Wednesday as the first guises enacted by washermen and weavers start their circuits through Old Tirupati, other activities begin at the Gangamma temples of Tallapaka and Tatayyagunta. Generally, donning guises during the Jatra is popular in Tirupati to fulfil a vow to Gangamma, to ask for protection, or as a pastime. During the first three days of the Jatra most of these guising participants are youngsters, mainly boys turned out especially as Bairagi ascetics, Ruffians and Sweepers. The Bairagi bodies are ash-coloured; those of the Ruffians split, with one side of the body red and the other green or black; while the Sweepers are blackish and carry little bottles of green 'liquor'. They flow raucously and full of fun through the streets and alleys in and around Old Tirupati, darting this way and that like schools of fish, ragging passersby, especially women.

They also go in droves to the courtyard of Tatayyagunta Temple to beat the stone feet of the Goddess. They say that they are 'fucking Gangamma' or 'fucking her cunt'. Groups of youngsters chant as they rhythmically beat her feet: 'Your cunt is as big as a winnowing basket, and the penis that doesn't fuck it is as small as an ant.' Mothers hold little sticks for their very small sons and together they beat the feet, while other women worship. Others spill salt and pepper on the feet. Coconuts are broken, the milk spilled over the feet; and here and there a rooster is beheaded and offered to the feet. Still other women adorn with turmeric and vermilion a high stone post embedded in the ground between Gangamma's feet and the entrance to her shrine. Of the beating and chanting, men say that Gangamma is full of desire (*kāma*) and they are giving her the sexual satisfaction she wants. Women say that after beheading the Palegadu, Gangamma turned into the stone head within the Tatayyagunta shrine. The abuse is intended to call her, entice her, out of her temple so that she will aid people.

The general atmosphere is of good-humoured exuberance. Similar emotions and rhythms of activity also predominate on Thursday. On Friday floods of little Sweepers swirl through Tatayyagunta. However, older teenagers also move through the temple courtyard, often in groups, beating Gangamma's feet and chanting abuse. But now the beating is done not with bare sticks but with branches of calming neem, and these branches are heaped on the feet and on every other ritual object in the vicinity. It is no coincidence that this attention to Gangamma's feet, though still calling her closer, is more modulated and calming on the day the female Sweeper appears in the streets with her winnowing basket, a clear sign that the Goddess is nearing the world of human beings. On Saturday the temples and their courtyards are quite empty of people. Gangamma's devotees await the killing of the Palegadu.

The Goddess Revealed

Before daybreak on Sunday the soon-to-be-revealed Goddess, her head covered, slips from the weavers' home into the night accompanied by a drummer and washerman. On his head the washerman carries the visiting middle-of-the-street Tallapaka Gangamma. They are on their way to Veshalamma.

The true form (*viśvarūpa*) of Gangamma is a single figure, the Matangi. She wears a sari of small red-and-white checks resembling a gingham pattern. Her entire head is wreathed with jasmine. Her garland is again half neem, half jasmine. Around the waist outside her sari is a folded cloth pouch filled with neem leaves and 'lap-rice' (*odibālu*), especially beneficial for fertility. The Matangi holds a small mirror attached to a handle,

9. Youngsters Beating Tatayyagunta Gangamma's Feet (Don Handelman).

10. Tatayyagunta Gangamma's Feet (Don Handelman).

cushioned with neem leaves, both taken from Gangamma's shrine in the weaver's home.

On reaching their destination, middle-of-the-street Tallapaka is placed before and to the side of the image of Veshalamma. The Matangi faces both. The images of middle-of-the-street Gangamma and Veshalamma are reflected in the Matangi's mirror. Looking at the Matangi from their perspective, their images are contained within her. The Matangi then performs the first ārati of the morning before the two, an act of worship normally done by the priest, but here emphasizing the unmediated contact among the goddesses.

There is here a profound moment of self-recognition in which these goddesses—middle and periphery, interior and exterior—fuse together within the Matangi. They see themselves within her, as she sees herself within them. As they recognize themselves within her, so does she within them. This is the ultimate self-union of the Goddess, of her coming into full self-awareness within the ritual cosmos of the Jatra. The Goddess is fully revealed to herself through their recognition of themselves in her. She receives her sense of self through them as the viśvarūpa uniting all of them.[15] The interior and exterior are identical, indistinguishable. A striking indication of this self-union is that on this day the Matangi, Veshalamma and Tallapaka Gangamma all wear the same sari of red-and-white checks. When the visiting middle-of-the-street Gangamma returns to the weavers' home, she too is offered a sari of the same pattern.[16] This union is made explicit both by ritual specialists and devotees. Thus on the first day of the Jatra a woman possessed by Gangamma at the Tallapaka shrine exclaimed: 'The two sisters will come together and I'll have a good Jatra.' In this instance the reference to self is to both the Goddess and the devotee.

These goddesses fuse on the border between the exterior and interior, between the self and the other, as the prince and his minister are killing the Palegadu further up the road. At this moment all the significant males of this ritual cosmos are beyond its border, the Palegadu dead, the Prince

[15] Here the mirror is a medium of reflexive holism and encompassment. See, for example, Beck's (1976: 241) reference to the use of the mirror in renewing the powers of a divine image. Among Newar, mirrors may be implicated in the representation of holism (Levy 1990: 682, 769). In other instances the mirror may be a medium of transitory transference (e.g. Bentor, n.d.). Also see note 8.

[16] On this day Tatayyagunta Gangamma wears a sari of reddish pink. Beck (1969: 558) considers the colour combination in south India of red and white to have auspicious, life-giving qualities.

11. Matangi: Home Visit (David Shulman).

and his minister cast aside, the male detritus of disguise and ritual action. The holistic Goddess, her head uncovered, begins her circuit from the ritual boundary, moving into Old Tirupati. As noted, much of the ritual movement in space is from the boundary to the middle, from outside to inside. This also maps the identity transformation of the Goddess from outside (the disguise of maleness) to inside (the revelation of the Goddess emerging from within). In the concluding section I will argue that this pattern of movement is linear, processual and progressive.

In writings on south Indian goddesses the mythic Matangi is described as an Untouchable (as is the Sweeper), the daughter of a sage and an Untouchable woman. She is married but never loses the power of her virginity. As a married virgin, she is a living-goddess (e.g. Hershman 1977: 273, on Bengal). In various village rituals her role was described as absorbing impurities through sexual relations or the touch of her buttocks. In village rituals, liquids spurting from her mouth purified persons of higher castes, including Brahmins (Elmore 1984: 25–6).

The Matangi's circuit takes over twenty-four hours. She is full of interior movement and energy, often guided by the elbow as if in danger of falling ever. She moves with a sense of purpose, walking briskly, accompanied by a drummer and during the night by a sweeper carrying an oil lamp to light the way. She is given vermilion continuously. Her hands, arms and sari turn red. Her lap-rice pouch, her womb, is turned from calming white to active red by the vermilion pressed on her. An ambivalent colour, red, in this context, most likely effuses health and vitality (Beck 1969: 558). In exchange, she blesses her worshippers by distributing lap-rice for fertility from her waist pouch and placing her hand on the heads of devotees.

The Matangi, the Goddess revealed, is the first of the sequence to enter within the domestic space of homes (including that of the weavers), in order to be worshipped there. The more the Goddess emerges from within herself, the more she enters the interior, intimate, 'middle' space of her devotees. She does not enter all homes, but stays within some for up to half an hour. The Goddess revealed is distant from the unmarried, violent goddesses described in south India who enter living areas of communities to be worshipped, but whose presence is feared (e.g. Good 1983; Reiniche 1987; Van den Hoek 1979; Moffat 1979: 261). Like the visiting Tallapaka Gangamma, the Matangi of Tirupati is more a middle-of-the-street Goddess, an intimate of inner domestic matters, a friend of the family. In a folk play published in Tirupati in 1915, Gangamma is described as coming to earth, as coming towards people, in order to protect them against fevers that do not originate from her (Kirusnapillai 1915).[17] Again her depiction is closer to a goddess of the middle. And on Sunday afternoon the now fully self-aware Matangi returns to the weaver household, once more face-to-face with the visiting Tallapaka Gangamma.

Before Tallapaka is a new sari of red-and-white checks. The Matangi sits opposite her in the narrow courtyard of the weavers' home. This small area is packed with people. Still seated, wreathed in garlands, her eyelids heavy, the Matangi's reddish body sways with energy, her bare feet stamping continuously to the drum beat. Suddenly she is up, approaching Gangamma, and sits again. The rhythm of the drum, the movements of her body and feet, are quicker. She is up once more, bouncing speedily forward towards Tallapaka with a jerky, hopping gait, tongue extended in fierceness. The Matangi falls on her side, is helped up, her tongue quickly pierced, given neem leaves to chew, and sits again. There should be no bleeding. Neem leaves are held in front of her lap, water poured over them and collected in containers on the floor. Those assembled now worship the Matangi continuously, and children are placed in her lap. The Matangi's tongue is pierced three more times during her circuit, once within the home of a Brahmin family (see also, Thurston 1909: 42).

[17] I am indebted to David Shulman for a summary of this work.

Tongue-piercing without blood-letting is a demonstration of Gangamma's power, but it also seems to calm her. For that matter it may well be both since she is the cosmic encompassment of the ritual cosmos, and therefore the source of benefice and self-control. Worshippers enhance her power with redness, and modulate it with water and neem. So, too, when she rushes towards the visiting Tallapaka Gangamma, tongue extended, her fierceness is modulated by the piercing. Immediately afterwards she is worshipped intensively, and she responds by distributing the rice of fertility and well-being. Without doubt the causal shaping of the Matangi's emergence from within the guises accomplishes one of the major goals of this ritual cycle—the bringing of goodness by the Goddess directly to devotees, without any external mediation.

The role of priestly mediation between the deity and worshippers is normative practice within temples and when deities emerge from their shrines and are processed through living areas. Whoever fills the priestly role accepts offerings on behalf of the deity, takes a portion for the latter and returns the remainder, the deity's leftovers, to be consumed or used otherwise by the worshipper. But as Gangamma's living presence swells day by day within Tirupati, she enters into direct relationships with her children, accepting their devotion and reciprocating with benefice.

Yet the presence of the holistic Goddess is difficult (and perhaps dangerous) for people to bear. Her next form is intended to leaven the powerful impact of the Matangi on worshippers. On Sunday afternoon the two lime pots (*sunnapukuṇḍalu*) appear. Once more these are the sisters Gangamma. Their facial markings are those of the Matangi. One's sari is saffron coloured, the other's saffron and maroon. In the left hand each holds a mirror like the Matangi's, and also vermilion that is spread on a bed of neem leaves. Attached to the hair of each is a white lime pot. The five cosmic directions are marked in vermilion on each pot. The pots have stood before the visiting Tallapaka Gangamma for some days, absorbing her power.

A latticework of sticks extends from the mouth of each pot. The pot itself is covered with tan-coloured cloth and tied in place on the head of the Goddess by cloth cords of the same colour. Strands of flowers—especially jasmine—are woven around the latticework. The pots are attached to the hair of the Goddesses as they sit in the courtyard of Veshalamma Temple, the crowd worshipping them. As they move, their flowered headdresses sway above the heads of those around, looking like miniature temple towers. Their circuit takes some forty-eight hours, with brief stops to empty monies offered by worshippers from the pots. During this period the Lime Pots are fed only fluids, especially buttermilk. In contrast to previous circuits that followed preset stops, the Lime Pots are accessible to any family along the route that desires their presence. Stops, however, are on the pavement outside the home. Their movement is slow. In some neighbourhoods, women from almost every home worship them. In contrast to the Matangi, the Lime Pots expand the penumbra of the worship of the revealed Goddess and the distribution of her benefice.

However, the oneness, the totality of the Matangi is fragmented into twoness, and this devolution from holism diminishes the power of the Goddess. The weavers say that the power of the Lime Pots lessens as they move through space and time because Gangamma is about to leave Tirupati. Whereas the holistic fusion of the Matangi erased divisions of the horizontal (those of the middle and periphery) and the vertical (those of transcendence and presence) within the ritual cosmos, the devolution of the Lime Pots reopens the horizontal division, since the sisters are Tallapaka (middle) and Tatayyagunta (periphery). The reappearance of divisions prepares ritual space for the disappearance of the Goddess from the city. The entire composition of the Lime Pots emphasizes the modulation of power. Their flowers and the colour of their saris are calming. The Lime Pots themselves are calming since they are thought to absorb impurities, just as lime absorbs heat.[18] To their worshippers the Goddesses distribute the life-giving redness of vermilion. Previously devotees had enhanced the power of the Matangi by giving her the vital redness of vermilion—now the Lime Pots divest themselves of some of this vitality, returning vermilion to worshippers.[19]

The Lime Pots are the last form of the Goddess to focus on the ordering and renewal of the domestic domain. The final appearance of the Goddess, the close of the Jatra, joins the domestic, ritual sequence of the weavers to rites at the Gangamma temples of Tatayyagunta and Tallapaka. After Saturday's lull, from Sunday to Tuesday, the kind and intensity of activity at the temples change, as they do in the city streets and in homes.

[18] At the end of the circuit the pots are broken at the spot that the Palegadu was killed. The pieces are divided among the weavers and drummers, and in water are thought good for fertility.

[19] The weavers have dropped one form of the Goddess—the female ascetic Jangamma—who had followed the Matangi. Jangamma's purpose was to modulate the impact of the Matangi's power. This function is now given over entirely to the Lime Pots. Their circuit is now lengthier, longer in time and more remunerative.

Multitudes of Gangamma

During the days of the revealed Goddess, children in guise disappear from the temple courtyards and the beating of Gangamma's feet ceases. Guising by teenagers and adults increases sharply; women perform rituals for Gangamma in their homes; parents perform rituals in temples to protect their children against disease; and an increasing number of families make blood offerings in the Tatayyagunta temple courtyard. Activity is frenetic, and numerous actors in different acts criss-cross one another in ongoing movement, occupying the same physical space with near simultaneity. The common focus is the presence of Gangamma and the hopes and desires that people have of her. Embodiments, reflections and echoes of Gangamma are seen and heard throughout Old Tirupati and beyond.

In many households, women dressed in their best saris worship Gangamma, often on Sunday, the time of the Matangi, and on Tuesday, that of the final appearance of the Goddess. In general, these are rituals to modulate Gangamma in relation to the home. Thus the Goddess may be consecrated as middle-of-the-house Gangamma (*naḍa iṇṭi Gaṅgamma*) in a pot containing a calming mixture of fermented rice water and millet covered with neem leaves, a flower for her hair on top, the pot placed on a yantra. After the women ingest Gangamma, the mixture is put outside and fed to passersby, thereby spreading the modulating effects from within the home. Gangamma is also consecrated by women in the home as a mound of clay with a flower on top, surrounded by neem leaves, placed on a new sari and fed a variety of modulating foods, some of which is again given to passersby.[20]

Family members take children to Tatayyagunta Temple to break 'thousand-eye' pots.[21] The reference to 'eyes' in the name of the pot may refer to the sores of the pox disease. Gangamma sometimes is said to have a thousand eyes; such sores may be called 'eyes' (Beck 1969: 572); and in a myth of the goddess Mariyamman, she herself equates a thousand pox sores with her thousand eyes (Beck 1981: 131). An older family member circumambulates the temple clockwise, holding the pot against the head of the child. At the very rear of the temple, facing the auspicious, life-giving east, the pot is hurled to the ground and smashed. Thus the child's

[20] I am indebted to Joyce Flueckiger for summaries of these rites.

[21] The contents of these pots commonly contain a camphor flame, two metal eyes, a black bangle and coins.

head is offered to the Goddess in the hope that she will reciprocate with a whole, renewed head.

Youngsters embody the Matangi in a modulated condition in order to worship the Goddess for protection against epidemic diseases, or to fulfil vows of thanksgiving for having survived such illnesses. Their heads are covered with calming yellow cloths heaped with braided jasmine; their garlands are mainly jasmine; and their clothes are in calming whites and yellows or fruitful greens. Covering their entire faces and sometimes their feet, over a base of soothing sandal paste, are the sores of the pox, painted in red, black and white. They travel in rickshaws or walk (sometimes preceded by a drummer to announce their passage) to Gangamma temples to show themselves to be in harmony with the Goddess.

Young men show themselves to be devotees of Gangamma by smearing their faces and upper bodies with sandal paste and, preceded by drummers, marching militantly in groups through the city and temple courtyards. Not a few adult men embody the Matangi by dressing as lovely, perfectly groomed and ornamented women in beautiful saris. The embodiments of many are utterly feminine and alluring. Others fulfil vows to the Goddess by taking on her various disguises or by dressing as characters from the epics. Through numerous embodiments by males the emergence of the female from within is shown on the surface of the self—the male made female by the female rising through the male.

Especially on Tuesday a profusion of family groups come to the Tatayyagunta courtyard to behead chickens and offer their blood to the Goddess. The earth is smothered by layers of feathers as the chickens are plucked and cooked. And the lines of people waiting to be seen by (and to see) the Goddess in her shrine lengthen. In contrast to the raucous good humour around the temples during the three days of the Goddess disguised, the last three days of the Goddess revealed are permeated by seriousness and solemnity, as numerous individuals and family groups seek their own ways of harmoniously aligning themselves and their needs with the presence of the Goddess. As Tuesday's late afternoon turns into night, knots of people swell into crowds that settle down to await the final appearance of the Goddess, early Wednesday morning.

Gangamma Meets Gangamma and Transcends Herself

By Tuesday evening a platform covered by a canopy is erected on the stone post between the entrance to Tatayyagunta's temple and her stone feet. This is also done at Tallapaka. Clay, mud and straw are piled to the

12. World Turning Female (Don Handelman).

side. After midnight the straw is twisted into rough rope and wound round the post above the platform. The making of the Big Goddesses has begun at each temple.

The final form of the revealed Goddess is called the Perantalu. Her name in Telugu seems to mean a 'big' or 'full' woman. Perhaps the gloss of 'complete' or 'whole' is appropriate. She leaves the weaver's home a couple of hours after midnight and reaches the finished Big Goddess at Tatayyagunta at daybreak, moving on quickly to the Big Goddess at Tallapaka. The figure of the Perantalu refers to a married woman whose husband is unable to consummate their marriage because he dies, and whose virginity is threatened by another male. Since only her husband should take her virginity, and since she is married, still a virgin but without a husband, she is a married virgin who also refuses to surrender her virginity to the alien aggressor. She retains her female wholeness, her power increases tremendously (e.g. Egnor 1980) and she becomes a deified virgin, a powerful Goddess transcending the human world (Narayana Rao 1986: 158–60; see also Hershman 1977: 273).[22]

[22] This a non-Brahminical version. The Brahminical Perantalu refers to an auspicious woman with a living husband. 'Perantam' refers to a gathering of such women for ritual purposes. In keeping with this, Elmore (1984: 26) comments that the deified Perantalu is an auspicious woman who has died before her husband. See also Thurston (1909: 42).

According to her myth, Gangamma destroyed the threat (the Palegadu) to her own virginity and has remained a virgin. She is perceived as such even with children; and she wears the tali, the wedding badge, though no one bothers about her marital status. For that matter, she has eliminated maleness as a power from her ritual cosmos. During her visit to Tirupati she has come closer to and is more intimate with human beings. Now she will take precisely the Perantalu trajectory from deified woman to transcendent cosmic encompassment—in departing from this world to a high, distant plane. She will leave in place the horizontal division between Goddesses reopened by the lime pots, and will expand this into the vertical separation between her ascendance into cosmic holism and her self-fragmentation through the Big Goddesses. However, the epistemological premises that existed prior to her arrival will not be exactly extant once more. The ritual causes a crucial shift in these premises that will be discussed in the concluding section.

The Perantalu's facial markings, adornment and dress are virtually indistinguishable from those of the Matangi. Her head decorations were used by previous forms of the revealed Goddess. She wears the red-and-white checked sari of the Matangi, and the lap-rice pouch at her waist is already stained red by the vermilion pressed on the Matangi. She is the direct continuation of the Matangi, but her trajectory is exactly the opposite. Instead of relating herself wholly to this world as did the Matangi, the Perantalu is about to cut her intimate connection with it and separate herself. In the process she takes apart the ritual structuring of her own Jatra.

With covered head, the Perantalu leaves in the direction of Veshalamma at about the time that the making of the Big Goddesses begins in earnest. She stops at the ritual border across from the Veshalamma temple. In the process of separating herself, she is not to come face-to-face with a part of herself (e.g. Veshalamma) who will remain in place in this world. The Perantalu removes her head cloth and performs a rite that dismantles the mandala made to keep her within the space of the ūru when her self-awareness there was minimal. Now she frees her own movement and then starts on her short circuit of stops, breaking off before daybreak to go to the Tatayyagunta and Tallapaka temples.

At these two temples, curtains are raised between the platforms and the assembled crowds. Behind each screen, clay mud is thrown at the straw base and smoothed and shaped into the form of a great female head over two metres in height. Features and ornamentation are added: silvered metal eyes, nose, ears, balls of hair, an open mouth and upper forehead both coloured deep red by vermilion, crown, forehead mark, nose ring, earrings, tali and garland. Her skin is coloured a deep purplish-black.[23] By 4 a.m. the Big Goddess at Tatayyagunta is completed, the curtain dropped and worshippers are offered vermilion from her mouth.

The Big Goddess is Gangamma, although the weavers deny that she is the Gangamma within the shrine. Nonetheless (echoing the women cited earlier in connection with the beating of Gangamma's feet), worshippers have worked to entice the Goddess to come out of her temple, closer to them and their needs. During the past few days they have attracted her attention through sexual play (beating her feet and verbal abuse), worshipped and fed her continuously as her hunger increases, decorated all signs of her presence with colours, scents, textures and designs that please her, offered her heads and blood, and taken her forms to show her to herself. Like the visiting Tallapaka in the weaver's home, Gangamma's presence at these temples swells; and for a brief period on this last night she emerges from her shrine towards her devotees, just as the Perantalu is about to leave Tirupati. These great heads of the Goddess are talked about as very fierce (*ugra*). Their eyes will burn all before them if they are not modulated. Yet a mother is overheard telling the small son in her arms to look directly into the eyes of the Goddess and not to take his eyes off her for a moment, hardly the inculcation of fear.

Within her shrine the Goddess's head faces east, as do her feet in the courtyard. In the form of the Big Goddess, Gangamma's head emerges in direct alignment with, and coming closer to, her stone feet. The height and depth (head and feet) of the Goddess are thereby embodied outside the shrine. Emerging from within her temple, she comes from deep within, from the 'middle' space that is also the height of the axis mundi. In so doing she moves in the directional axis of west to east, from a 'higher' positioning (west) to a 'lower' one (east) (Beck 1976: 215–16). Her movement towards her worshippers and her embodiment there is a descent from her shrine in an auspicious direction (east). Appropriately, by worshipping her feet her devotees will receive her head.

All await the coming of the Perantalu together with the first rays of the rising sun. The Perantalu is said to tear out the eye and cheek of the Big Goddess (destroying the power of the fierce eye). On her arrival, the

23 The heads differ in some details. Tallapaka's head is somewhar smaller; her mouth is less protuberant and her crown is higher. Wooden fangs are stuck into Tatayyagunta's mouth, but not in Tallapaka's.

13. Composing Big Goddess at Talapaka Gangamma Temple (David Shulman).

Perantalu stretches up to touch the mouth of the Big Goddess and the crowd immediately tears apart her massive head, distributing bits of clay mud amongst themselves. Within two or three minutes nothing remains of the Big Goddess apart from the straw core. Drunk in water or put in the earth, the clay is good for fertility.

The Perantalu departs swiftly to fragment the Big Goddess at Tallapaka. In facing this Big Goddess, the Perantalu once again points roughly west, the direction of height or superiority in the east-west axis of cosmic force (Beck 1976: 215). From Tallapaka she continues in a westerly direction, the direction of rain (Beck 1976: 233), spiralling inwards within the mandala. Near its centre she rests from her labours for some minutes at a pavilion that faces west and disappears from there into the waters beyond the seven seas. The Perantalu's circuit takes her round the periphery of the ritual space that she is taking apart—past Veshalamma and the fragmentation of the Big Goddesses—transcending all these forms of herself, and ascending from the mandala's middle.[24] Two days later rain should fall, cleansing the impurities generated by the mixing together of castes during

[24] One is reminded here of the three strides of the Vedic Visnu, in which the third is the highest, ascending through the cosmic centre to the cosmic height (Kuiper 1962: 146–8).

14. Perantalu (Joyce Flueckiger).

the Jatra. A few weeks later the vital monsoon season should start in earnest, proof that Gangamma indeed was fully present in Tirupati.

Through the Perantalu the Goddess eliminates herself from the earthly plane. In this cosmos made female by the Goddess, female cancels out female to end the ritual cycles of the Jatra—that of Tallapaka Gangamma's visit to the weavers' home and those of rites at Gangamma temples and in homes. In the moment of separating herself from her full presence of self on the earthly plane, the Perantalu comes face-to-face with her self made earthly, the Big Goddess. In terms of the sequencing of revealed forms of the Goddess, the Big Goddesses are similar to the lime pots. There are two of them; they re-establish the division of the middle (Tallapaka) and periphery (Tatayyagunta), but they also bind this division to particular places, the temples; and they are fully present among their worshippers. Thus the Big Goddesses are akin to a revealed form of the Goddess that appeared earlier in the sequence, that the Perantalu has superseded. The Perantalu is, as one Weaver put it, equal to and greater than the Big Goddesses. I interpret this in terms of the temporal progression of the Goddess's revealed forms. The Big Goddesses are equal to the Perantalu, but at a time that has already progressed beyond itself, towards the separation from Tirupati of the revealed Gangamma. Therefore in the present time of their face-to-face encounter, the Perantalu is later and greater, moving once more toward transcendence and the holism of cosmic encompassment.

The Goddess stretches herself cosmically: simultaneously she fragments and devolves her self (the Big Goddess) and transcends and evolves her self. The process is highly transactional, in keeping with the ontic maxim of Hinduism that to give is to get. In her self-fragmentation she gives herself in small bits to the multitude of her devotees, entering into them, their homes, their soil, just as they have given themselves to her in myriad ways during the past days. At the same moment she disappears and ascends. Momentarily, as she stretches herself, her force is diffused, and perhaps her devotees experience her separation from the human plane as the opening of space suddenly revealed within the confining connectivities of cosmos, as a spaciousness for new beginnings (e.g. Halbfass 1992: 30–2).

Gender, Cosmos, Caste

I commented at the outset that the Gangamma Jatra raises questions about the formation of identity among south Indian goddesses. A good many scholars have attended to these women through studies of myth and ritual, but on the whole the results are disappointing, reproducing over and again the distinction between unattached, violent women associated with peripheries like ritual boundaries, pacific women domesticated by their husbands, and the conditions under which one is turned into the other. This may well reflect a prominent, male south Indian perspective on goddesses, perhaps one held by women as well. Yet this ideology certainly does not preclude significant counter-formations of gender identity that are no less embedded in and empowered by tradition.

Table 1 The Days of the Gangamma Jatra and the Sequence of Guises

Tuesday: Middle-of-the-street Tallapaka Gangamma comes to the weavers' home.
Wednesday morning: The ritual mandala is made to keep Gangamma within the ūru.

The Goddess Disguised

Day	Guise	Enacted by	Comments
Wednesday morning	Male Bairagi ascetic	Unmarried washerman	Announces the Jatra
Wednesday afternoon	Female and male snake-charmers	Unmarried weaver and washerman	Worshipped in street at set homes
Thursday morning	Female and male shepherds	Unmarried weaver and washerman	Worshipped in street at set homes
Thursday afternoon	Female and male ruffians	Unmarried weaver and washerman	Worshipped as above; female's nose is cut off and she is unreflexive to this disfigurement
Friday morning	Female and male merchants	Unmarried weaver and washerman	Worshipped in street at set homes
Friday afternoon	Female and male sweepers	Unmarried weaver and washerman	Worshipped as above; clamourings for female's blessings with winnowing basket; female prominent; Goddess more self-aware
Saturday afternoon	Male prince and male minister	Married weaver and washerman	Worshipped as above but more intensively; do not go to Veshalamma; breach ritual border; kill Palegadu and eliminate maleness

Table 1 (*cont.*)

The Goddess Revealed			
Day	Guise	Enacted by	Comments
Sunday	Matangi	Married weaver	Self-revelation and self-recognition of Goddess; union of Goddesses at Veshalamma; gives lap-rice; worshipped as above but also enters homes
Monday afternoon	Lime pots	Married weavers	Modulating impact of Matangi
Wednesday morning	Perantalu	Married weaver	Opens mandala and ends ritual cycle; distributes herself by fragmenting
			Big Goddesses and holistically transcends earthly plane

Where should one go for these contesting conceptions of gender? Ritual is one useful possibility, since in traditional societies its logics often do not reflect in simple ways the organization of social life, as so many scholars continue to maintain (Handelman 1990: 1–20). Contrary to commonsensical scholarship (e.g. Bloch 1989), numerous rituals are stores of cultural imaginings and help in modelling and re-ordering social life. Performing rituals is the practice of these imaginings. In India this has probably been prevalent since Vedic times. Rites of south Indian goddesses are no exception. They probably have their own cosmic metaphysics of gender formation and identity that is latent, yet present (and therefore active), in the world of human beings.

The Gangamma Jatra is dominated by men embodying the Goddess in public places. The bulk of the devotees are women. From this perspective it appears that at the core of the Goddess there is the ideology of male perception and male control that is worshipped by females. If so, then this is one more study that reproduces this ideology. Nevertheless I believe that the epistemology of gender in this Jatra is more complex, and that this is related directly to the kind of cosmos that Gangamma engenders. The metaphysics of goddess identity may well evoke cultural imaginings of gender that are shut down in social life. Discussions of the metaphysics of Indian gender should acknowledge first and foremost the premises of lability, fluidity, and processuality that pervade Indian cosmological imaginings (see Chapter Two and Handelman, 1992a). Hindu cosmology is organized through transformations that are continuous and flowing, and not through discrete, sharp-edged separations that are akin to, for example, binary oppositions.

In their temples, the models of their cosmos, south Indian deities of black stone live deep in the interior, in the dimly lit cavelike centre, the middle that is simultaneously the height of the axis mundi. Regardless of temple size and wealth, or the popularity of the deity, each Goddess and God in principle encompasses the cosmos. Each is therefore the cosmic whole, existing simultaneously in the middle (the whole) and high up (the whole), and further out and lower down (as parts of that cosmic whole, and therefore as parts of itself). Moreover, the relationship between the whole and part is in continuous movement and flux (e.g. Chapter Two in this volume), the deity moving out and down, in and up. This too is the movement of the deity in relation to the worshipper, moving down and out to come closer, in and up to go further away.

This kind of cosmos is integrated systemically by connections (*bandhu*) between everything and everything else. The further in, the higher up the deity moves, the closer it comes to self-realization as the cosmic encompassment, the more its self loses otherness and self-awareness. The epistemology of holistic encompassment at the highest plane of the cosmos suggests that the deity simultaneously exists everywhere and everywhen. Thus the deity knows everything, because everything is connected. And at this highest level, everything is connected simultaneously. The cosmic self of the deity is then the cosmos. There is no other. This level of encompassment, of total knowledge, is constituted through the extreme density of its connections. Given the simultaneity of knowledge and consciousness, there is no space, no place and no time, since each of these is premised on separations and gaps in the fabric of connectivity. Therefore the further in and higher up the deity moves, the denser the connectivity of the cosmos, and the more dense the deity becomes.

The thickness, solidity, impermeability and seamlessness of stone approximates this density of the cosmos. The cosmos implodes. The deity's turning into a rock on the human plane is an index of this great interiority and distance from human beings. It is an index too, not of stasis, but of the simultaneity of movement that is total knowledge, and that on the human plane may appear as a kind of 'frozen' motion, the interior movement of rock that has nowhere but everywhere to move, so long as all movement is simultaneous and not sequential. This density of holistic encompassment is the epistemic index of cosmic hierarchy. Concomitantly, the descending, devolving movement of the deity opens tears and separations in the connectivity of self, cosmos and knowledge, introducing the godhead to time and space, and lessening the density of the cosmos and deity on lower planes. This emphasis on the significance of thickness and multiplexity in the constitution of the cosmos echoes the importance of density and quality in human relationships of the everyday (e.g. Daniels 1984; Trawick 1990).

How does this formulation of the cosmos shape the self of godhood? At the plane of total encompassment there is self, but no other. Therefore there is no self-awareness. People and deities share the profound transactionalism of the Indian world (e.g. Marriott 1989). Thus self-recognition exists through the perspective of otherness. In the *Ṛg Veda* the first moment of self-recognition by the Cosmic Self (Brahman) that has no Other is profoundly paradoxical (Handelman 1992a: 7). It is also the beginning of the differentiation and evolution of the cosmos (Miller 1985: 53–8). In Hindu mythology the impulse to self-awareness on the part of the deity often requires an external stimulus, almost a deus ex machina (as in the instance of drawing forth a separate Śiva and Parvati from the androgynous Śiva, the Ardhanarisvara). The cosmic process is the movement of the self. The first motions of self-awareness by the deity (that look to us like movement because they are through time, linear and sequential) begin centristically, within and high up. These movements generate gaps in the cosmic fabric of connectivity, and therefore doubts about the nature of reality and its ambiguities. As David Shulman comments: 'In any Hindu context, movement along the axis mundi is also an internal process within consciousness' (1998: 211).

The shaping of the deity's self, of increasingly reflexive self-awareness, is integral to self-identity, and for that matter to the conscious empowerment of self. The deity acquires self-awareness by coming out and down towards devotees. The deity is simultaneously the cosmic encompassment and a part of itself, a fragment. This part is the active deity who as a lower, less dense form of itself synchronizes with worshippers and shares with them experiences of time and space. Time and space exist because of the separation between the cosmic encompassment and its part, the opening of gaps in the density of the cosmos on its lower planes. The problem for the worshipper on the human plane is how to induce or seduce the deity into her embodied self and therefore into otherness, thereby paradoxically parting with herself as she moves into conscious selfhood and empowering self-awareness.

These formulations speak of the emergence of the self and the shaping of identity in the disguise and revelation of Gangamma. Every disguised and revealed female form of the Goddess is empowered within the weavers' home by Middle-of-the-Street Gangamma, but most show themselves by removing their head coverings at Veshalamma's peripherality. The exceptions are the prince and his minister who are addressed further on, and the Perantalu who takes apart the ritual structure. Each form emerges from deep within the middle (of the Goddess, the home) but is externalized at the border (of the Goddess) and (with changes in form) moves progressively deeper within the living space and being of others, her worshippers. This movement has a linear, processual character: from within the Goddess (herself as her own middle) outwards to the interface of her self with others (herself as her own border), and then further out, into the lives of those others. This movement is also progressive: on successive appearances the self of the Goddess is a little more shaped, a little more present in this world.

Before the Jatra begins the Goddess is elevated deep within herself, far away in the stony density of cosmic encompassment. In this condition she has little awareness of herself and others, and so no perspective towards herself. In other words, as the cosmic encompassment she neither knows nor has any need to know who she is. To acquire self-knowledge she needs an external perspective that will tell her who she is and how she is different from others. When she goes from her self-centredness (Middle-of-the-Street Tallapaka Gangamma temple) to the weavers' home, she is far away, unaware of who she is. The mandala keeps her within the ūru of Tirupati since she might wander away accidentally—as yet she is without self-knowledge of her purpose there. Yet as noted, even the mandala cannot keep her from remaining deep within herself, or for that matter from returning there. Therefore she must be seduced out of and into herself. To do this, her embodiment in time and space is crucial. She is embodied in an active, fierce, red-coloured form who carries a sword (the visiting Tallapaka) and has a male head (the Palegadu) next to her. But despite the transposition from her shrine, this new locus is still one of 'middle', within the interiority of the weavers' home. In epistemological terms, to know herself she will have to leave the interiority of middleness (her own, the weavers' home) in order to return to the interiority of others (her devotees) with the self-awareness to recognize the reciprocity between her self and theirs, and therefore the benefice she has for them.

Therefore she is embodied further in a guise and given a self (someone who she is not), given a husband (who is not), taken to the border of herself (Veshalamma, the Mother of Guises, the interface of interior and exterior), exteriorized (she removes her head covering) as who she is not (the surface self of the guise) and put on the street. Dressing her as a snake-charmer, her devotees tacitly put the question: 'Is this really you?' She wanders from place to place as a snake-charmer with a snake-charmer husband, passing the exteriors of homes. Within this prosaic, prosthetic self she comes into direct, unmediated contact with others. Yet they do not treat her as a snake-charmer. Coming as a snake-charmer she is received as the Goddess. People wash, decorate and touch her feet, and offer her fire and food. They worship her, they love her, the lowly snake-charmer, as the Goddess. Who then is she? These householders deny the authenticity of her exterior self and show her who she really is in the human world. Each guise given her—shepherdess, ruffianess, merchant—is denied publicly as her true self. The ruffianess disguise—her meek acceptance of disfigurement by her husband—shows that she is still far away, unaware of her true self. Nonetheless, each guise is responded to in the same way. Regardless of her changing self-presentations, she evokes the identical response of respectful, devotional worship. The external perspective toward her is inherently stable, telling her over and again that she is not her exterior. Then who indeed is she? This external perspective insists consistently that she is the Goddess. Is she attracted enough, curious enough, to come out and find out?

Worshippers expect little in return from the Goddess in these early disguises. Instead they labour to give her appreciation and love, to attract her attention and awaken her emotions. They introduce the Goddess to female embodiment, itself intimate and full of feeling. Through the sequencing of guises, they take her from the simultaneity of infinity and synchronize her with human time, divisive, fragmenting, limiting. As she moves through the streets she experiences the separations of space and place. Her worshippers are persuading her to fragment, to externalize, to come to her own surface of self, to shatter exterior forms that are superficial and transitory, to recognize how very attractive to them she is, and so should they be to her. From the Goddess's point of view, these guises are journeys of self-discovery. The closer she comes to the surface of the human world and the more she embodies human attributes, the more attractive these are.

The embodiment of the female sweeper with the winnowing basket signals to worshippers that the process of making the Goddess self-aware is working.[25] This is the first time in the sequence that she is treated as someone with a consciousness of self, and therefore with the power to deliberately choose to give benefices of serious value. As one who is aware of herself, she is asked to reciprocate what she is receiving. She is entreated, implored, blocked, pushed and pulled into clapping heads with her basket. Everyone on the street, householders and passersby alike, insist on receiving her blessings. Their mood is joyful and demanding. Her pseudo-husband is almost ignored. The importance of the female Goddess swelling within the female guise is now apparent. This is the last appearance in which the female is paired with a male. Males have a significant place in this ritual cosmos only as long as disguise does. Males, one may say, are a part of the disguise of the Goddess as a meekly married woman. As the absence of her self-awareness gives way to self-revelation, so maleness dissolves into superfluity.

The disguise of the Prince (following immediately on the Sweeper) marks the violent eruption of the Goddess into a recognition of self. She knows who she is but cannot exfoliate her self fully. At each stop outside a home her minister enunciates explicitly what the Goddess will do to the Palegadu. The full self-awareness of the female is blocked by the male, by a male-dominated social order, perhaps by the male self insisting that there is nothing of the female within him. She has done away with the disguise of husband-wife pairs that normatively subordinates woman to man. She is now the male for the first time, the Prince at the pinnacle of male power, superior to his companion, the minister. Yet full self-awareness, the ability to take an external perspective on herself, requires that she externalize herself into the human world. To do this she must recognize that as a female in this world she is constrained by maleness that she must break through. The remaining block to her full self-recognition is the epistemological male whose own self-awareness depends upon his normative control and dominance over the female, upon rigid gender divisions, and upon his monopoly over power and authority.

The Palegadu perceives the surface as reality, seeing only himself in the warrior prince. His maleness blocks the Goddess's emergence into the

25 The cultural logic of guising in this ritual sequence is systemic and causal. The appearance of the female sweeper works as a feedback loop to show the participants that the cosmos is responding to the design of the rites by producing the correct effects (See Handelman 1990).

human world because he refuses to recognize the female within himself, and therefore the resonance that he too has with her on a deeper, interior level of being. He sees a male-dominated cosmos, and is closed to any external perspective towards himself that is not empowered by maleness. The Goddess uses the appearance of maleness to eliminate the male who perceives only appearance. To do this, the Jatra creates a ritual cosmos that on its surface subordinates women to men, wives to husbands, and that has only males at the apex of the hierarchy—the Prince, his Minister, the Palegadu. Then everyone is shown that she is fully conscious of herself, as she deliberately and violently tears this fabrication to shreds. The Prince and Minister are Goddesses; the Palegadu is dead, not to be revived since he will be just as he was; the female has embodied the male and made him part of herself, and the cosmos is female.

The revealed Goddess is perhaps the female in everyone. She unites with other significant females of these rites—the visiting Tallapaka Gangamma and Veshalamma—and deliberately reciprocates worshippers for their efforts to awaken her hunger to their devotion in this world. Emerging from deep within herself, surfacing through the exterior of the guise, she leaves exterior public space to enter deep within domestic intimacy where her true form is worshipped. In a sense she holds everything together in its interiority, just as the mother holds everything together at the hearth. The Matangi is the culmination of the causal design of the Jatra, intended to make Gangamma descend and emerge, active and energizing in the Tirupati ūru. Analogous processes at the temples of Tallapaka and Tatayyagunta crest there in the emergence of the Big Goddesses. As the Perantalu leaves Tirupati she fragments her lower self (the Big Goddesses) that makes its way in bits and pieces of labile mud into the bodies, homes and soil of her devotees, while she ascends into the density of cosmic encompassment, of no-time and no-space.[26]

---

[26] David Shulman's essay on the myths of Venkateśvara (n.d.b: 14) demonstrates that in the broader cosmology of Tirupati, 'male and female are inside-out images of one another'. Venkateśvara too is given to guises of self. Yet, whereas Gangamma emerges from her stony self-encompassment, incorporating maleness on the way, Venkateśvara's guises lead him deeper into himself, losing himself within the density of stone and incorporating femaleness on the way. Thus the female and male are cosmic complementarities in a profoundly generative sense—as the female (Gangamma) incorporates the male on the human plane, the male who is said to have been female (Venkatesvara) encompasses the female on the cosmic plane. One would expect that as the female (Gangamma) returns deep within herself, encompassing maleness, the male will emerge from deep within himself into the human plane. Though speculative for now, one should note that Govinda Raja (Visnu) ritually emerges from his temple into the world at the end of June, following Gangamma's

Yet what manner of cosmic density is this? Is it once again the stony density of connectivity from which Gangamma emerged and took shape? Not likely. The Jatra effects an epistemic shift in the nature of cosmic encompassment and its detritus. Gangamma disappears into the waters beyond the seven seas. The highest plane of her cosmos is now water. How is water different from stone? Water is no less dense than stone in its inner connectivity and simultaneity of time and space. Nonetheless water is, one may say, a different medium of cosmic density, one that flows and that changes shape and positioning within itself. The fluidity of Gangamma's epistemic shift is something akin to the stretching of connectivity without changing its density—thereby creating a sense of roominess. The movement from the inner density of stone to that of water loosens internal shape to generate spaciousness within the intense connectivities of no-time and no-space.

Gangamma returns within herself together with, at least momentarily, the holistic fullness of self-awareness that is also her encompassment of and relation to the other, her devotees. This self-recognition that fully encompasses and relates to the other, this spaciousness within extreme density, is akin to love and compassion for the self and the other. Of interest here is the comment that among Tamils, rain and coolness are associated with ideas and feelings of love and compassion (Egnor 1984: 26). The epistemic shift from stone to water is one from encompassment without awareness of self to encompassment with that recognition. In the latter instance the self-consciousness of the cosmos extends to the plane of human beings. The fluidity of water signifies this fullness, and so too the conscious penetration of cosmic benefice into this world. The descent of water fills the momentary spaciousness within density with the flowing compassion of Gangamma, with the fall of rain.

Earlier I suggested that there are deep fractures between metaphysical formations of gender and its social construction. Engenderment in Hindu cosmology is frequently flexible, flowing, paradoxical and self-transformative. Gender in Indian social life is often hierarchical, divisive and rigid. The metaphysics of gender identity open cultural imaginings that are shut down in the gender of social roles. The Gangamma Jatra explores contradictions of this sort, and shows that they too are disguises, concealing the deeper unity, or perhaps the unified flow of gender. At issue is whether

return into herself, and coinciding with the emergence and descent of her benefice, the monsoon rains. Venkatesvara's emergence follows a couple of months later.

the metaphysics of gender in south India point to a continuum of gender identities or to their hard-and-fast division. The Gangamma Jatra implies that ontologically, gender is a continuum; but that epistemologically, the experiencing of gender as a continuous flow surfaces only in specialized contexts.[27]

The premises of gender at the outset of the Jatra are the following: the source of benefice is female; the female (Gangamma) does not know herself; the male (Palegadu) perceives his own nature as utterly masculine and as absolutely superior to the inferior female; and so the male forces from the female what is of value to her (her virginity, her inner state of wholeness). During the causal course of the Jatra, these premises are transformed. The female (Gangamma) emerges into herself through the male (the weavers) disguised as a female (the guise). In layers, the qualities of both the male and the female are shown to be more superficial than the deeper character of the female, the Goddess who encompasses maleness. The emergence of the female from within herself moves her through a boundary that is both male and female.

The male as a boundary is of two kinds—the male (weavers) through whom the Goddess moves towards her own exterior, and the male (Palegadu) who blocks her movement because he sees nothing but the homogeneity of maleness. She destroys completely the male who blocks her movement towards self-awareness. She incorporates into her femaleness the male through whom she moves towards herself. The crucial moment is when the Matangi revives the Palegadu and shows him what a woman really is. What indeed does she show him? My guess is that she reveals to him what he adamantly refuses to see—the female in every male, and therefore the male within her self. She shows him gender as a continuum of cultural imaginings. In the meantime the Goddess has enabled the male she moves through (the weavers) to experience the

---

[27] On steps, however hesitant, towards the metaphysical imagining of gender as continuum, see Belo (1949: 14–17) on Java; Trawick (1990: 73, 252–3) on south India; Bynum (1991) on medieval Europe, and especially Strathern (1988) on Melanesia. In this regard, useful contrasts may be established with, for example, ancient Greek imaginings of gender. See Tyrrell (1984: 88–112), duBois (1982: 110–28) and Detienne (1989). On imaginings of gender in one Western locale of the fantastic, see Handelman (1991, 1992b). Though in this essay I conceptualize gender as a continuum, a more accurate metaphysical depiction would be spheroid. On the surface of the sphere, aspects of gender meet in their numerous differences and combinations. As the layers of the sphere peel off, onion-like, the differences of gender become less and less distinct. At the 'core' of the sphere, its middle and apex, there are no gender differences, no distinction between self and other.

female in his interior. The creation of a female cosmos does not eliminate the male perspective—but demands the blending of perceptions of gender in cultural imaginings.

The female as a boundary is herself—Veshalamma, the Mother of Guises, the border between the interior and exterior of self. As Gangamma surfaces into full self-awareness, she unites with Veshalamma, erasing the border between interior and exterior, middle and boundary, female and male. During this time the social division of gender moves closer towards the metaphysical unity of gender. And males throughout the city who appear in public as well-groomed, beautiful women are experiencing the continuous, metaphysical unity of gender, the female within their being.

If gender is a continuum, then this has implications for the design of ritual and cosmos. In India each gender is often linked to different powers that are made manifest and causal through specialized contexts of ritual. A signal problem of ritual design, therefore, is to construct the kind of cosmos that will draw forth from the gender continuum that gender (or aspects of this) whose powers are needed. In this section I argued that through ritual the self of the deity is made and shaped. Now I should add that from the perspective of ritual design, rite makes gender, and remakes it continuously from rite to rite, in accordance with its goals and intentions. As cosmic selves are made, so is gender. The Gangamma Jatra makes and shapes femaleness to tap its powers, keeping the male out of this cosmic construction. The engenderment of this cosmic self is an especially powerful mode of being and ideological integration for the peoples of castes termed left-hand.

The division of castes into right-hand and left-hand has been integral to Telugu society (Narayana Rao 1986: 142), as it has among neighbouring Tamils (Beck 1973; Mines 1984: 146–7). Much of the discussion on south Indian goddesses locates them among right-hand agriculturalist and landowning castes. Narayana Rao argues that in the epics of these castes the men 'aspire to heroic warrior status and keep their women under strict control' (Narayana Rao 1986: 147). So too in narratives of these castes, women turn their men into heroes, sending them to defend the borders, while they themselves fade into domestic obscurity. However these also are the social groups who often worship unattached, disease-bringing goddesses of the peripheries whose ethos is distinct from goddesses of the centre. Thus the right-hand configuration of goddess worship contrasts clearly between the meekness and placidity of the woman under her warrior husband's control and the violence of the peripheral woman.

The epics of left-hand, trading and artisan castes are woman-centred: the female heroically defends the integrity of the social group against an alien, aggressive power (Narayana Rao 1989: 114). Among these castes, contends Narayana Rao, 'women represent an inner strength and men remain largely passive' (1986: 147). In Tirupati all the castes that have some formal role in the Gangamma Jatra are of the left-hand. Moreover all the disguises of the concealed Goddess are either of left-hand castes or of tribal peoples with a leftish ethos (e.g. mobility in space). Gangamma worship in Tirupati generally, and particularly in the Jatra, shows a pattern of goddess organization that is distinct from that prevalent among right-hand castes. Gangamma worship posits the same goddess at the centre and periphery, with both sharing numerous attributes. In the centre is the domestic (but not domesticated) middle-of-the-street goddess whose presence in the midst of the community is desired, who has children but whose marital status is of no especial consequence, and who is a source of benefice. At the periphery is the unattached goddess who has children (if by force), who is more threatening but who also protects against disease, and whose marital status again is not an issue. Both are guardian figures. Both collaborate in defending the ūru against the incursions of the threatening, alien male, and both together kill him. Moreover these Goddesses unequivocally unite into one, rather than substituting for, or alternating with, one another. The united Goddess is then the source of benefice, of fertility and protection. In rituals of left-hand castes, as in their epics, the position of women jointly as agents of domestic well-being and territorial guardianship seems distinct from that of women in right-hand castes. Nonetheless the role of goddess rites and their cosmologies in the integration of left-hand castes has hardly been broached.

The weavers of Tirupati say that Gangamma's *śakti* (her female force) split into four: one married Visnu, one married Śiva, one married Brahma and one fragmented into 365 parts, each of which became an ūru. Tirupati is one of these. Cosmologically, the ūrus of Gangamma fill the calendrical solar year and are spread correspondingly through space. Metaphysically the integrity of the Tirupati ūru is woven into the role of the female as protectress both of the domestic domain and of the entire social unit. Epistemologically this agency of the female may depend on male recognition in ritual of the fluidity of gender that enables men not simply to impersonate female protectresses, but to become them. The idea of gender as a continuum may have especial resonance with left-hand caste integration.

Bibliography

Babb, Lawrence A., 1970. 'Marriage and Malevolence: The Uses of Sexual Opposition in a Hindu Pantheon', *Ethnology*, 9: 137–48.

Beck, Brenda E. F., 1969. 'Colour and Heat in South Indian Ritual', *Man* (n.s.); 4: 553–72.

——, 1973. 'The Right-Left Division of South Indian Society', in *Right & Left: Essays on Dual Symbolic Classification*, Rodney Needham (ed.), Chicago: University of Chicago Press, pp. 391–426.

——, 1976. 'The Symbolic Merger of Body, Space and Cosmos in Hindu Tamil Nadu', *Contributions to Indian Sociology* (n.s.), 10: 213–43.

——, 1981. 'The Goddess and the Demon: A Local South Indian Festival and its Social Context', *Puruṣārtha*, 5: 83–136.

Belo, Jane, 1949. *Bali: Rangda and Barong*, Seattle: University of Washington Press.

Bentor, Yael (n.d.). 'On the Symbolism of the Mirror in Indo-Tibetan Consecration Rituals'.

Bloch, Maurice, 1989. 'Symbols, Song, Dance and Features of Articulation: Is Religion an Extreme Form of Traditional Authority?' in his *Ritual, History and Power*, London: Athlone Press, pp. 19–45.

Bradford, Nicholas J., 1983. 'Transgenderism and the Cult of Yellamma: Heat, Sex and Sickness in South Indian Ritual', *Journal of Anthropological Research*, 39: 307–22.

Bynum, Caroline Walker, 1991. 'The Body of Christ in the Later Middle Ages: A Reply to Leo Steinberg', in her *Fragmentation and Redemption: Essays on Gender and the Human Body in Medieval Religion*, New York: Zone Books, pp. 79–117.

Champakalakshmi, R., 1981. *Vaisnava Iconography in the Tamil Country*, New Delhi: Orient Longmans.

Cox, Arthur, 1881. *Manual of the North Arcot Gazetteer*, Madras: 326–7.

Daniels, E. Valentine, 1984. *Fluid Signs: Being a Person the Tamil Way*, Berkeley: University of California Press.

Detienne, Marcel, 1989. 'The Violence of Wellborn Ladies: Women in the Thesmophoria', in *The Cuisine of Sacrifice Among the Greeks*, Marcel Detienne and Jean-Pierre Vernant (eds), Chicago: University of Chicago Press, pp. 129–47.

duBois, Page, 1982. *Centaurs and Amazons: Women and the Pre-History of the Great Chain of Being*, Ann Arbor: University of Michigan Press.

Egnor, Margaret, 1980. 'On the Meaning of Śakti to Women in Tamil Nadu', in *The Powers of Tamil Women*, Susan S. Wadley (ed.), Syracuse: Maxwell School of Citizenship and Public Affairs, Syracuse University, pp. 1–34.

——, 1984. 'The Changed Mother or What the Smallpox Goddess Did When There Was No More Smallpox', *Contributions to Asian Studies*, 18: 24–45.

Elmore, W. T., 1984. *Dravidian Gods in Modern Hinduism*, New Delhi: Asian Educational Services.

Fuller, C. J., 1980. 'The Divine Couple's Relationship in a South Indian Temple: Mīnākṣī and Sundareśvara at Madurai', *History of Religions*, 19: 321–48.

Good, Anthony, 1983. 'A Symbolic Type and its Transformations: The Case of South Indian Ponkal', *Contributions to Indian Sociology* (n.s.), 17: 223–44.

Halbfass, Wilhelm, 1992. *On Being and What There Is: Classical Vaiśeṣika and the History of Indian Ontology*, Albany: State University of New York Press.

Hanchett, Suzanne, 1988. *Coloured Rice: Symbolic Structure in Hindu Family Festivals*, Delhi: Hindustan Publishing Corporation.

Handelman, Don, 1984. 'Inside-out, Outside-in: Concealment and Revelation in Newfoundland Christmas Mumming', in *Text, Play and Story: The Construction and Reconstruction of Self and Society*, Edward M. Bruner (ed.), Washington, D.C.: American Ethnological Society, pp. 247–77.

——, 1990. *Models and Mirrors: Towards an Anthropology of Public Events*, Cambridge: Cambridge University Press.
——, 1991. 'Symbolic Types, the Body and Circus', *Semiotica*, 85: 205–25.
——, 1992a. 'Passages to Play: Paradox and Process', *Play and Culture*, 5: 1–19.
——, 1992b. 'One Leg in the Air: The Iconography of Engenderment in Western Circus Posters', Symposium on Genres, Genders, Methods: The Symbolic Economy of Circus in Art, Literature and Ethnography, Victoria University, Toronto, June 5–7.
Hershman, Paul, 1977. 'Virgin and Mother', in *Symbols and Sentiments: Cross-Cultural Studies in Symbolism*, Ian Lewis (ed.), London: Academic Press, pp. 269–92.
Kiruṣṇapiḷḷai, Vikkiravāṇṭicālai, 1915. *Keṅkaiyammaṉ Nāṭakālaṅkāram*, Tirupati: Irattiṉa Mutaliyār and sons (in Tamil).
Kolenda, Pauline, 1981. 'Pox and the Terror of Childlessness: Images and Ideas of the Smallpox Goddess in a North Indian Village', in her *Caste, Cult and Hierarchy*, Meerut: Folklore Institute, pp. 198–221.
Kuiper, F. B. J., 1962. 'The Three Strides of Viṣṇu', in *Indological Studies in Honor of W. Norman Brown*, E. Bender (ed.), New Haven: American Oriental Society, pp. 137–51.
Lannoy, Richard, 1971. *The Speaking Tree: A Study of Indian Culture and Society*, New York: Oxford University Press.
Levy, Robert I., with the collaboration of Kedar Raj Rajopadhyaya, 1990. *Mesocosm: Hinduism and the Organization of a Traditional Newar City in Nepal*, Chicago: University of Chicago Press.
Marriott, McKim, 1989. 'Constructing an Indian Ethnosociology', *Contributions to Indian Sociology* (n.s.), 23: 1–39.
Miller, J., 1985. *The Vision of Cosmic Order in the Vedas*, London: Routledge & Kegan Paul.
Mines, Mattison, 1984. *The Warrior Merchants: Textiles, Trade, and Territory in South India*, Cambridge: Cambridge University Press.
Moffat, Michael, 1979. *An Untouchable Community in South India: Structure and Consensus*, Princeton: Princeton University Press.
Moreno, Manuel and McKim Marriott, 1989. 'Humoural Transactions in Two Tamil Cults: Murukan and Mariyamman', *Contributions to Indian Sociology* (n.s.), 23: 149–67.
Narayana Rao, Velcheru, 1986. 'Epics and Ideologies: Six Telugu Folk Epics', in *Another Harmony: New Essays on the Folklore of India*, S. H. Blackburn and A. K. Ramanujan (eds), Berkeley: University of California Press, pp. 131–64.
——, 1989. 'Tricking the Goddess: Cowherd Kāṭamarāju and the Goddess Ganga in the Telugu Folk Epic', in *Criminal Gods and Demon Devotees*, Alf Hiltebeitel (ed.), Albany: State University of New York Press, pp. 105–21.
Oppert, Gustav, 1978. *On the Original Inhabitants of Bharatavarṣa or India*, New York: Arno Press.
Ramanujan, A. K., 1986. 'Two Realms of Kannada Folklore', in *Another Harmony: New Essays on the Folklore of India*, S. H. Blackburn and A. K. Ramanujan (eds), Berkeley: University of California Press, pp. 41–75.
Reddy, K. Viswanadha, 1976. 'Ganga Jathara—A Folk Festival in a Small City—Tirupati and the Neighboring Villages in Andhra Pradesh', *Folklore*, 17: 237–42.
Reddy, V. Narayana and A. Munirathnam Reddy, 1991. 'History and Patronage of Tirumala-Tirupati Devasthanam: An Anthropological Study of the Temples of Tirumala and Tirupati', in *Social Anthropology of Pilgrimage*, Makhan Jha (ed.), New Delhi: Inter-India Publications, pp. 141–76.
Reiniche, M. L., 1987. 'Worship of Kaliyamman in Some Tamil Villages: The Sacrifice of the Warrior-Weavers', in *Religion and Society in South India*, V. Sudarsen, J. Prakash Reddy and M. Suryanarayana (eds), Delhi: B.R. Publishing Company, pp. 89–105.
Shulman, David D., 1980. *Tamil Temple Myths: Sacrifice and Divine Marriage in the South Indian Śaiva Tradition*, Princeton: Princeton University Press.

Shulman, David, 1998, 'Ambivalence and Longing: Vyāsa's Curse on Kāśī', in B. Z. Kedar and R. J. Zwi Werblowsky, *Sacred Space: Shrine, City, Land*, London and Jerusalem: MacMillan and Israel Academy of Sciences and Humanities, pp. 192–214.

——, n.d.a., 'On Being a Stone: A Reading of the Tirupati Purāṇas', Ms.

Sontheimer, Günther-Dietz, 1989. *Pastoral Deities in Western India*, New York: Oxford University Press.

Thurston, E., 1909. *Castes and Tribes of Southern India*, vol. III, Madras: Government Press.

Trawick, Margaret, 1990. *Notes on love in a Tamil Family*, Berkeley: University of California Press.

Tyrrell, W. Blake, 1984. *Amazons: A Study in Athenian Myth Making*, Baltimore: Johns Hopkins University Press.

Van Den Hoek, A. W., 1979. 'The Goddess of the Northern Gate: Cellattamman as the "Divine Warrior" of Madurai', in *Asie du Sud: Traditions et Changements*, Paris: Editions du Centre National de la Recherche Scientifique.

Wadley, Susan S., 1980. 'Sitala: the Cool One', *Asian Folklore Studies*, 39: 33–62.

CHAPTER FOUR

GROWING A KINGDOM: THE GODDESS OF DEPTH IN VIZIANAGARAM

Don Handelman, M. V. Krishnayya and David Shulman

The priest sits carefully in a makeshift wooden seat attached to the very end of a fifty-foot pole, a tamarind tree with its bark and branches removed, rubbed golden yellow with turmeric (*pasupu*). On his lap, wrapped in silk, are three elongated pieces of wood. A fourth, slim and attached to a squarish rectangle, he clutches firmly in one fist. The pole swivels against its upright wooden fulcrum, itself set into a wheeled carriage. Dressed as a king, wearing a royal turban, the priest is tied into the seat with new saris. Self-contained, reserved, he is surrounded by police who shape the great crowds into a semblance of orderly flow. At the signal, as the crowds shout and pelt the priest/king with bananas, hundreds of thousands of them, the pole is raised high at a forty-five degree angle to its carriage. With its precariously dangling figure ensconced on his seat, the tree top reaches forward, into the sky, as the carriage begins its trundling journey down the packed, narrow thoroughfare toward the Kota, the Fort, the palace of the king, some hundreds of yards away. The tree is the goddess, Paiditalli, the Golden Lady, who some two and a half centuries earlier was the younger sister of Peda Vijaya Rama Raju, the king of Vizianagaram. As she carries the king aloft, Paiditalli begins the *Sirimanu Jatra* ('Goddess Tree Movement'), the climactic event of some five months of ritual acts that, in our words, grow the kingdom of Vizianagaram (and its kingship). Before Indian Independence, Vizianagaram was a little kingdom in northeastern Andhra Pradesh, in the region called Southern Kalinga; the Sirimanu Jatra continues, perhaps because growing the kingdom was related intimately to the growth rhythms of cosmos, and cosmos continues.

Growing kingdom and kingship is our problematic: How is this done? What kind of cosmos potentiates the growing of kingdom? What are the implications of growing cosmic kingship for the little kingdom in this region of India? How does the growing of kingdom resonate with the growing of rice? With the growing of the Goddess? And, within all such questions, though we can touch on this theme only briefly in this essay, what is the nature of the Telugu tie between sister and brother, one through which the male is himself shaped and formed by the female with whom he has an organic relationship? To grow a kingdom and its kingship quickly comes to imply growing a cosmos in all its complexities, with 'growth' not a metaphor for these dense connectivities but rather a continuous dynamic through which cosmos re-forms itself within itself as it proceeds through itself, spatially, temporally. Our immanent concern is to explore this cosmos through its Goddess, its ritual, and its history.

1. *Prologue: Every King Needs a Goddess. On the Curved and the Flat, Wilderness and Civilization*

Every Andhra king needs a goddess, but not every Andhra goddess needs a king. Moreover, a king needs a goddess who is a sister to him, comforting, consoling, protecting, and saving him without asking questions, without setting conditions. The goddess who is (like) a sister to a king is a goddess to kingdom and kingship precisely because she is (like) a sister. Kingship needs nurturance; it needs to grow, organically, from depth into depth, from depth out of depth, like a field of rice. And like a rice field, kingdom and kingship need to be fed, watered, and protected from mishap, assault, and disease. Like rice, then, kingdom and kingship must find the rhythms of being planted, transplanted, grown, harvested. The goddess of kingship and kingdom, Paiditalli, grows these rhythms, and she herself grows into these rhythms as she grows them. Therefore she herself is grown or, more precisely, she grows herself together with the cosmos that is her kingdom. As the goddess grows, so does kingship. This too is the nature of the sister/brother tie—a linkage of deep organic affinities (unlike that between husband and wife, more a forced conjunction of opposites). This means that, whatever they do, sister and brother naturally activate, affect, and effect one another, as do goddess and king. The yearly growth of the Goddess is that of the self-organization, the autopoiesis, of her Vizianagaram cosmos.

We argue that in the little kingdom of Vizianagaram, kingdom and kingship are grown organically, like a crop of rice. Together with rice, kingdom and kingship grow into maturity—rice the sustenance for human beings, the others the organizers of sustenance, yet themselves sustenance for cosmos and goddess. Kingdom and kingship are grown during the annual growth cycle of rice, dependent on the great southwest monsoon of June–July. The monsoon rains that grow rice are brought by the Goddess,

Paiditalli. Simultaneously, during and after the monsoon season, kingdom and kingship are grown by the Goddess as she herself, once the younger sister of the king, grows and matures. Perceived in this way, kingdom and kingship become a created and creative endeavor, made fertile within the growing cosmos of the goddess and growing together with, indeed through, cosmic rhythms. Sirimanu jatras are widespread in the farming communities of this region of Andhra, though they usually are held in January and seem exclusively focused on agriculture. In Vizianagaram the Sirimanu was adapted to kingship by binding together the kingdom, the rice-cycle upon which sovereignty depended, and the bond of brother and sister, a protective and nurturing nexus of Gajapati kingship. We will have more to say on the Sirimanu and kingship in the conclusion.

Like the rice field, kingdom and kingship need to be made fertile, watered, transplanted, grown, harvested. Both rice field and kingdom must be opened into depth, deepened into the spaces within which the creative impulse is generated, intensified, germinated. This is the doing of Paiditalli, who opens and deepens cosmic space within which the kingdom is prepared and shaped and within which kingdom and king are joined to one another. These dynamic depths are crucial for creative growth, whether of plants or kings, Tantric Yogis, wrestlers, poets, or Ayurvedic healers—all of whom flourished in Vizianagaram in the not distant past. Paiditalli is, we may say, a goddess of depths. Yet Paiditalli herself must be persuaded to appear within her cosmos, within the shallower surfaces inhabited by human beings and, once there, to grow herself as this cosmos, to take root deep within it, as her presence matures here. As the goddess grows, transforming surface into depth, so are kingdom and kingship prepared for growth; and as kingdom grows, so does rice, the sustenance of kingdom.

In this cosmos, growth of any kind occurs only within depth, emerging from inside, from within the interiority of interiority. The space of interior depth enables the forming of form, the intensities of forming that no less make form different in and from itself (Deleuze 1994: 252). Concomitantly, an increase in the intensities of qualities of form changes that form from within itself. Such change is basic to growth.

Growth needs space/time in which to grow; and, appositely, the energy of growth, of increase, the intensity of its qualities, opens space/time into its own evolution. Depth is fluid, labile, and in depth there is the fluidity that is transformation, the fluidity of forming and forms that take shape through one another. *Change is the condition of depth.* What is often referred to as '*wilderness*' in India, supposedly on the peripheries of '*civilization*,' is where depth exists in its natural cosmic condition. Put differently, wilderness is a greater concentration and intensification of depth. In or close to regions of greater depth—seas, lakes, streams, and forested mountains full of caves, whose fluidity or inward-turning of form curves them more naturally—are the abodes of depth-specialists: fishers, hunters, healers, and of course, goddesses. Depth-specialists are experts in transformation, entering depths and there joining with the interior dynamics of cosmic process. Depth specialists modulate the intensities of qualities of form, thereby altering these. The dynamics of cosmic process are interior—it is there, through deep and pervasive flowing, that form is forming, that growing is begun and form begins before it emerges onto the surface.

On the surface are kingship, kingdom, cultivated fields, all identified with the more rectilinear, stable, centered building of civilization by males. These are *flatter* dimensions of existence. They are flattened and hardened, one might say, by the gravity of sociocentric hierarchy, by its more hard-and-fast social categories, by its complex divisions of space and labor. So, too, fields that are not prepared for fertile growth will flatten, collapsing into themselves and compacting, since without growth they have no depth; they do not reach upwards, as it were, into depth. They will dry out, losing whatever interiority they possess, becoming increasingly dense, static, stuck, overly stable and therefore sterile. If fields are opened into the depths of furrows and into the heights between furrows, sown with seed, filled with water as it rains and flows from the depths above and is raised from the depths below (by well irrigation), then we can see the dynamics of growth and florescence. This is true for kingdom and kingship as well.

Goddess, waters, mountains, rice fields, kingdom, and kingship continuously grow into one another. Their relationships are deeply recursive, as the entirety of cosmos continues to develop and takes shape, folding into itself through them, enclosing itself within itself. This curvature is profoundly *fractal*, since the curves of these folds create themselves through themselves and therefore create themselves as similar or, perhaps, as harmonically resonating with one another, thus as fitting with, indeed into, one another.[1] In this regard, the natural locales of depth (in wilderness, wherever this appears) and the agents of depth-making and growth (in civilization) are never really distant topologically from one another,

---

[1] This joining of fractal and fold is influenced by Deleuzian thinking, developed through analyses of ritual. See Handelman 2005.

even though they may be separated from one another in topographically linear terms. As cosmos curls into itself, these fractal-like relationships are embedded dynamically in every ongoing moment and portion of the curvature. We will argue that there is no principled cosmological-topological distinction between height (of mountain, of sky, of furrow edge) and depth (of bodies of water—streams, lakes, seas, and the puddles between furrows).[2] Moreover, the mountains in this area of Andhra are porous, full of caves and tunnels, so that mountains, like water, can be said to be full of interior movement. Both mountain and water are deep, and the depth of a mountain fits into the depth of a lake, while the interior of a mountain (with its swirling caves and twisting tunnels) can take in the sea.

Consider the great mountaintop temple of Simhacalam, close to Vizianagaram, the abode of the god Simhadri Appana (Vishnu as Varaha-Narasimha, the combined Boar and Man-Lion), where the last Raja of Vizianagram, P. V. G. Raju, died as a poor pilgrim. The *sthalapurana* of the temple, its myth of origin, describes its mountainous heights as a 'mountain of rain within the ocean.'[3] Were cosmos to be entirely folded into itself, mountain would fit into water, and water and lake into mountain. The same could be said for sister and brother—for the sister who makes her brother deep enough to fit into her own depths, thereby enjoining her care and protection. In some sense, this occurs during the climactic moments of the Sirimanu Jatra.

The city of Vizianagaram is positioned on flatland between a set of craggy mountains holed with caves and a smallish lake (called Big Lake)—that is, between wilderness of height and wilderness of depth.[4] The mountains descend to the fields and the town and then farther down into the lake. The depth of the lake mirrors the depth of the mountains reaching into the sky. Turn the spheroid cosmos on its head, and the waters of the

---

[2] Our references to topology are not to the formal mathematics of space but rather to what may be called 'relational dynamics' of space/time in the ways in which cosmologies are organized. For example, some scientists think of time that, as topological mapping, is sometimes referred to as 'rubber sheet geometry' (Assad 1999: 41), in which time curls, folds, twists, so that any point of time may meet any other.

[3] According to one story on the place of origin of the Simhacalam temple, Simhadri Appana hopped from one mountaintop to another. looking for one strong enough on which to build his temple. In finding the right one, his feet went deep into its mountaintop and he could not extricate himself. There, Simhacalam was built. So these mountains too have inner space, pores, perhaps with mud and ooze within them.

[4] Even in the height of the dry, hot season the lake never dries up, providing irrigation to a considerable area and supplying numerous wells (Francis 1907: 336).

lake will fall on human beings and flatlands like rain, as rainwater tumbles down the mountains into the fields. Moreover, such resonant reflections occur no less in the tiny scales of cosmos, in the harmonics of scale among variant topologies.[5] Thus in the plowed field (unlike the dry flatland), the space between one furrow and another is a high mountain, while the furrow itself filled with water is a deep lake (as is the well). The topological dynamics of irrigation and the coming of the monsoon are fractal harmonics of one another.[6] Such harmonics are deeply present in the rituals of Paiditalli and her cosmic re-creation.

And when the lake waters ripple? When fluid turbulence twists on itself, when the wave becomes a higher fluidity, a mountain, the mountain becoming the outerness of wave, a crusted fluidity, stone full of fluid space? Water and mountain, height and depth, are distinctive features of Vizianagaram topology, intimately and dynamically related, perhaps turning into one another, continuous with one another. The convex curve of mountain turns into the concave curve of sea with its own convex curve of ripple and wave (that together are like the furrows of a plowed field). Such is wilderness in Vizianagaram. It is from such a dimension that cultivation, kingship, and kingdom come, through the goddess who dwells within the depths of wilderness, deep within herself, yet who emerges to form herself in order to form cultivation, kingship, state.

The continually rounding cosmos of water and mountain, below and above, thus differs sharply in one respect from the flattening cosmos of kingship and kingdom. The spheroid cosmos is fully continuous with itself, within itself, without sharply delineated interior boundaries or other hard-and-fast distinctions. This is why the depth of seas and lakes and that of mountains reaching into the heavens curve into one another. For that matter, *depth depends on continuousness*, on an ongoing deepening, depth going deeper, going elsewhere, yet also curving back through itself so that mountains enter sea and lake, just as water inundates the heights during the rains only to flow down into itself. In this regard, wilderness inevitably

---

[5] We eschew the commonly accepted distinction between macrocosmos and microcosmos in favor of a notion of harmonic resonances among different scales of cosmos that, in a dynamic sense, enable these scales to enter into and to become one another with continuity. The Vizianagaram cosmos of Paiditalli is one of immanence rather than transcendence.

[6] The dynamics of interiority are at one with themselves, emerging from, intensifying, and shaping phenomenal forms that existentially are these dynamics. In other words, we deny any significance in the Vizianagaram cosmos to the commonly accepted academic distinctions between form and content, structure and process, and the like.

surrounds the cultivated domain of fields, of human beings, because wilderness curves and is thus continuous, unlike the discontinuity of the cultivated, of kingdom constituted through neat divisions and separations. Little kingdoms in this region of South India are dots in a wilderness.

This continuousness of cosmos is identified with topological formations outside of civilization and cultivation, yet deeper than these (so that they appear within civilization as its drying and flattening). The denizens of wilderness often differ from those of civilization and kingdom. People of the wilderness enter into and are specialists in the flowing, recursive depths of cosmos. They are fishermen who plumb the currents and depths of water, like those who, as we will see, aid in the creation of the goddess Paiditalli. They are hunters who roam the steep heights and depths of mountains, about whom myths abound. They are healers who plumb the deep interiors of human beings (and deities). Fishermen, hunters, healers are integral to the holism of the cosmos; they *safeguard* this holism by continuously mending discontinuities as these appear. Stated differently, the continuousness of cosmos that is identified with wilderness makes these people continuous also with boundaries between wilderness and civilization. Their plumbing of depths situates them within boundariness and enables them to influence and control the bounded domains. This faculty, together with their affinity with the depth, is what makes possible their mending of rupture, their activation of the movement necessary to the well-being of cosmos. Ammavaru, Mother Paiditalli, has a special feeling for such people, for the Scheduled Castes (Dalits), the Madigas, for the Besta and Jalari fishermen, the Talaiyar oil pressers, for the bards and drummers, for the Uppara well-diggers, basket makers, and fix-it men. These are the ones who, as they bring goddesses from the wilderness surrounding villages into their middle space, also bound and bind the village, regulating it so that no other goddesses (*ammas*) or cholera and pox can penetrate and enter. So, too, these people are responsible for bringing Paiditalli from her watery depths, mythically and, to a degree, annually; they have the task of carrying her around the city and they join in her leaving.

Paiditalli, too, is a healer of kingdom and kingship, a healer who quickens time and opens space. During the dry season, when she is distant from humankind, disease flattens human beings, sickening and killing them. In the searing heat and absence of water, the earth shrinks and atrophies; growth is slowed and stunted. The world of human beings slows down, flattens even further into itself, becoming discontinuous with itself. Paiditalli brings rain, ending the hot season, opening depth and accelerating time for fruitful growth, shaping the world as whole again in its fullness. Paiditalli is the healer who rescues the human world from self-sterility.

Kings, on the other hand, venture into wilderness at their peril. Kingship grows out of cosmos; it does not impose itself on cosmos. Yet without the continual transposition of wilderness into discontinuity, civilization will not exist. With discontinuity, time slows, its fluidity enchained, marking stops and starts at innumerable boundaries.[7] Civilization depends on discontinuities that generate the organization of clean-cut categories, divided and separated from one another in order to be connected through limited operations or functions (of governance, of taxation, of the rule of law, and so on) that can be turned on and off. Rectilinear borders, regardless of their permeability, block the dynamic fluidity of movement. Therefore such borders also dry out, flatten and compact whatever they enclose. Boundaries order and regulate, making whatever they contain distinctly different from whatever is outside—more fixed in place, more congealed, more static. Thus kingship, kingdom, and civilization exist because order is created through discontinuity, through hierarchy and status, through distinctly demarcated caste groups and their living areas, through controlled cultivation. Kingship is the apex of the force and power of social discontinuity. Despite its surface powers, kingship is denied depth and its corresponding fluidities, and thus kingship may be perceived as flat and superficial in its topologies. Therefore every king needs a goddess to heal and rejuvenate him.

These relationships are no less temporal than they are spatial. Within the kingdom, time slows the continuousness of cosmos by setting it in distinct places marked by borders and division. Time is divided (by royalty, law, caste rights, land rights, and so forth) and thereby slows down. By contrast, wilderness, entering the kingdom, accelerates temporal dynamics.[8] The slowing of time has something to do with the flatness of

---

[7] One classical view of time, beautifully articulated by Bhartrihari (fifth century) and exemplified in Kalidasa's great essay on kingship, the *Raghuvamsa*, sees temporality itself as constituted by two countervailing aspects—a blockage or bunching, *pratibandha*, and a opening or release, *abhyanujna*. Far from an even, steady flow of identical Newtonian units, this kind of time progresses snail-like, with differential intensities, each occlusion eventually giving way to break-through, as drought gives way to flooding. One of Bhartrihari's images to convey this movement is that of the waterclock and its small, tenuous opening (*nalika-vivara*), through which water flows unevenly. See Shulman (in press) and Nidbach 2013; Bhartrihari, *Kala-samuddesa* 70.

[8] Wilderness may generate temporality. The ancient Vedic year was conceived as two months of monsoon and ten months of growth. These two months were a time

kingdom, the acceleration of time—its maturation or ripening—with the depth of wilderness. We might describe this contrast in terms of "dimensionality": depth has greater potentialities for movement, for formation from within itself, than does flatness. As time accelerates, there is, as it were, more exploration of the potentialities that are depth. This is the creative power of depth. The traversing of flatness is more lineal, more enumerative (as in measures of land and population, of property taxes, of the layout of a field, of a city, of a house), emphasizing the intersection of angular surfaces. Yet the traversing of depth has many forms that take shape through involutions, through odd routes and formations (like the movement of Paiditalli as the Sirimanu tree).

In South India, longing and desire seem to be primarily a function of fullness, not of emptiness or absence. It is fullness that has no bounds, that is moving through its intensities, expanding, opening space and time for the growth of greater desire.[9] In this sense, fullness makes fullness, especially, perhaps, in a watery world (what is fuller in its density than fluid?) that is depth on all sides of flatness. Fullness never empties but rather expands further *into* its own fullness. In Vizianagaram, the fullness that is depth is the desire and longing for growth and creativity whose primary medium in the kingdom is Paiditalli. In the climactic Sirimanu Jatra, fullness is *felt* tangibly in all its dimensions.

Nonetheless, kingship is responsible for the well-being of the kingdom. Therefore, every king needs a goddess through whom to reenter the fluid fullness of cosmos, perhaps to save kingship from itself, from its overweening drive for power and conquest. (The Gajapati kings of Vizianagaram called themselves *Manne Sultan*, the Lords of the Wilderness.) The integrity of kingship depends upon its relationships with wilderness beings—those who can mend and heal the cosmos—and above all with goddesses and, therefore, with the brother-sister bond, a tragic yet triumphant force in the myths of Paiditalli's origin.

---

of non-differentiation—between earth and water, sky and land and, in a sense, height and depth—that may be likened to wilderness, from which emerged the ten months of growth.

[9] This sense of fullness seems to relate to cosmos integrated from within itself, as we argued for South Indian Saiva cosmos (Handelman and Shulman 2004). Cosmos integrated from within itself seems not to be encompassed from outside itself [by God], and, as such, not to have external boundaries or limits. The absence of external limits speaks to how such a cosmos is integrated from within itself through its intra-relatedness [rather than its inter-relatedness]. See also Handelman and Lindquist 2011: 23–35.

2. *The Birth of Paiditalli*

Who is Paiditalli? How did she become the present-day goddess? How did she become so intimately related to wilderness and enmeshed in kingship? There are various versions of her origins, their appearance sensitive to context and to the identity of the narrator. The two accounts of immediate relevance here are related to the trauma of the destruction of Vizianagaram kingship and to its resurrection. These are no less accounts of the organic relationship between sister and brother. We first present the base narrative that is most directly related to kingship[10] and, following this, aspects of the account obtained from Bairagi Nayudu, Paiditalli's chief priest in Vizianagaram at the time of our fieldwork and a servant of this goddess since 1976 until his death in January 2010. Bairagi Nayudu's account highlights the bond between sister and brother. Historical and cultural information relevant to these accounts is discussed in detail in the following section.

In 1757 the Vizianagaram ruler, Vijaya Rama Raju, together with French and Hyderabadi troops, attacked the neighboring little kingdom of Bobbili, destroying the earth-walled palace-fort and its defenders who had assembled there in a last-ditch stand. Seeking revenge, the Bobbili hero, Tandra Papaya, slipped into the French camp at night and killed Vijaya Rama Raju.[11] Vijaya Rama Raju had a younger, unmarried sister named Paidimamba.[12] Paidimamba was a devotee (*daiva-bhakturalu*) of Vijayavada Durga (Kanaka-Durga, the Golden Durga), and was only fourteen, a generic age in South India for girls on the near edge of puberty, still virginal, though their marriage has been arranged. Vijaya Rama Raju and Paidimamba were motherless and fatherless. They had only each other[13] and Pativada Appala Nayudu, who was both bodyguard and minister; they

---

[10] This version is current in Vizianagaram both orally and in various printed versions, e.g. the 1994 brochure entitled *Mrokkina varamul' icce callani talli sri sri sri paiditalli ammavari katha caritra*, available in the Square Temple (pp. 5–6).

[11] See discussion of the historiographical sources in Narayana Rao, Shulman, and Subrahmanyam 2001: 24–92.

[12] To a degree we are assigning categories to the names of the Goddess—Paiditalli, Paidimamba—somewhat arbitrarily, here and elsewhere, in order to ease description. These and other of her names are largely interchangeable, without fixed referents or boundaries. The most prevalent way of referring to this goddess by her devotees is to call her Ammavaru, Our Mother. Naming the Goddess has no consistent usage that we could discover. She is also called Paidiyamma or Cinnamma (Little Mother).

[13] By the time of the Bobbili war, however, Vijaya Rama Raju was married, though childless.

15. Bairagi Nayudu (David Shulman).

grew up under his guidance and protection. Appala Nayudu was like a foster-parent to Paidimamba—when she was small she sat on his lap and he carried her on his shoulders. (Bairagi Nayudu understands Appala Nayudu as his ancestor.) Paidimamba had warned her brother not to go to war with Bobbili, yet he had paid her no heed. As Bairagi Nayudu commented, 'She was, after all, only a small child, not old enough to advise her brother. He merely smiled at her words.'[14] After the king left for Bobbili, the Goddess of the Vizianagaram kings, Kanaka Durga (Golden Durga), came to Paidimamba in a dream, telling her what was to befall her brother. (Some say that the dream came to Paidimamba while she was sick with smallpox, *maśūci.*)[15] Determined to prevent this, she rode off for Bobbili, accompanied by Pativada Appala Nayudu. As Paidimamba rode past Big Lake (*Pedda Ceruvu*) on the outskirts of Vizianagaram, news arrived of her brother's death. Paidimamba told Appala Nayudu that she was about to become one with the Goddess (Durga), to 'lose herself' in the Goddess, but that she would return, and that then her image should be worshiped on the other side of the lake (that is, away from the city, in the wilderness). She flung herself into Big Lake and drowned.

Some time after her death, she came in a dream to a low-caste Jalari fisherman, telling him to raise her image from the bottom of Big Lake. Jalaris dove into the water and found her image in the muddy bottom.[16] Untouchable Madigas and Malas recognized her image and took her in procession around the city. The image was established in the Wilderness Temple (*vanam gudi*), the original temple of the Goddess, across the lake from Vizianagaram. From then on, Paidimamba, the younger sister of the dead king, became the Goddess, Paiditalli, the protectress of the kingdom. The two were one. Paiditalli did have another shrine, a tiny one leaning against three tightly intertwined trees of three different species (Banyan, Neem, and Fig) growing together virtually as one not far from the Kota,

---

[14] Vijaya Rama Raju's anger at the king of Bobbili resulted from a cockfight that the former lost. The cockfight in Andhra, very much like a Deleuzian singularity, explains the sudden occurrence of a phenomenon for which there is no explanation. In this regard, the micro-dynamics of a cockfight (like the beating of butterfly wings in Chaos Theory) have unpredictable, probably destructive cosmic consequences (see Roghair 1982).

[15] Paiditalli ammavari katha caritra, 6. Note that smallpox is classically the disease of the goddess in south India.

[16] In this narrative, Jalari is used as a generic term for fisherman. The Jalari are sea fishers living on the eastern coast, north and south of the major port of Visakhapatnam (see Nuckolls 1996). The Besta of Vizianagaram are fresh water fishers (today, they are mainly tailors by caste occupation) and say that they were the ones who took Paiditalli's image from Big Lake. The Besta participate as a prominent group in the Sirimanu Jatra.

the royal palace-fort within the city. Paiditalli, it was said, would sit on a square mat beneath these tree(s). The small area of the trees is called *Balalayam*, the Temple of the Young Female. Indeed, it may well be that these trees are no less the Goddess. In 1924, during digging in this area, an image of the Goddess was unearthed and installed in a temple, the *caduru gudi*, the Square or Courtly Temple, the temple of the square mat (*caduru*), redolent with referents of royalty,[17] built in front of the tree and its tiny shrine.[18] The annual re-birth of the Goddess begins in the vicinity of the *vanamgudi* and shifts to the *cadurugudi*.[19]

Kingship collapses in its very expansiveness, in its outward-moving predatory character that includes the king's inattention to the sage advice of his younger sister to remain within the depths of interiority of the Goddess-sheathed kingdom. However, the movement of the king could be characterized as 'flat.' He moves in a linear mode from Vizianagaram to Bobbili, and this trajectory turns into the vector of his destruction. The sister-brother tie is clearly related to the difference between female and male; it situates itself somewhere between the vectors of conservation and fragmentation, between a curvature that strengthens the interior flows of cosmos and a shallow linearity that moves along feebler surfaces. No less important is the concomitant contrast between the potential for fertility (Paidimamba is an unmarried virgin) and the deadness of sterility (apparently the king dies without offspring). The desire of the male, the brother, is to move outward into destruction; while the female, his sister, counsels inwardness. Within the depths of interiority is the protection and salvation of kingship, ruined through the shallowness of exteriority. When Paidimamba receives the news of her brother's death out there in the kingdom of Bobbili, as foretold to her by the Goddess, she throws

---

[17] 'Square' likely also refers to the *mandala* according to which Vizianagaram (like so many other cities in South Asia) was built.

[18] Ramalingesvara Swami, an Ayurvedic doctor and close friend of Anand, the present rajah-who-would-have-been, connects the Square Temple to Indira Gandhi (despite the discrepancy in dates). One late night, on his way home from his clinic, he saw a car pull up in front of the tiny shrine of Paidimamba. His friend the last Raja got out, followed by Indira Gandhi who had just lost an election. Later he asked the Raja about what had transpired. The Raja had told Indira Gandhi that if she wanted to win the next election, she would have to do something for Paidimamba. So Indira prayed to Paidimamba and built her the new temple we see today. A midnight visit by Indira Gandhi to Square Temple (in the year in which the Congress Party split) was confirmed by Bairagi Nayudu who was introduced to her there by the last raja.

[19] Both temples are those of the Kapu (agriculturalist) caste. Paiditalli's priest, Bairagi Nayudu, was a Kapu.

herself into the lake, into wilderness, into depth, into interiority. Her death is undoubtedly tragic—the sister's self-destruction because of the needless loss of her beloved brother—but is it suicide, as an untempered response might suggest?

Suicide, after all, is usually understood as the end of self, the intentional destruction of self-interiority. Yet Paidimamba has announced that her intention is to unify with the Goddess and to return. Therefore her death should be understood as transformative, indeed as self-transformative through a self-sacrifice in which Paidimamba gives up her own particular selfness to become integral with Durga, the Goddess of Vizianagaram kingship. Goddesses, we are arguing, are denizens of wilderness, while wilderness is integral to the deep interior holism that *is* a goddess. Paidimamba, as a woman, is linked to wilderness and civilization. Her femaleness funnels her into the state of a goddess and, therefore, into wilderness; while her kin status relates her to kingship. Yet her kin status in relation to kingship is demoted on two counts: she is both female and a younger sister. Her self-sacrificial death ruptures the ambiguous relations among wilderness, femaleness, and kingship. Becoming Paiditalli, she is transformed into the protectress of kingship. The younger sister to the king, junior and ignored, is now, in a sense, an older sister to kingship, impossible to ignore. Through this transformation of sister into goddess, kingship is woven more and more tightly into the interiority of the goddess, into her cosmos, from within which kingship emerges and grows. The deaths of the king and his sister generate the re-creation of kingdom as space that should open into depths of creation (the locus of the goddess) from which growth emerges, with deep roots and deep heights (the cosmos of the goddess).

Most significantly, the transformation of Paidimamba occurs within wholly fluid depths. The lake is a natural container of cool fluid depths opening smoothly in all directions, perhaps a womb of the goddess within which Paidimamba moves into Paiditalli and is reborn as the unity of both. The king dies in his flat trajectory skimming shallowly along the earth. The fate of the flat, whenever its thinness is pressured, is to fragment; it is brittle, with minimal resilience. The sister, by way of contrast, unites with the goddess in the cool depths where the goddess is continuous with herself. It is in the violence of death that Paidimamba inflicts upon herself that she discovers depth, and this may imply the discovery of self-cultivation, the discovery of depth-in-selfness by entering wilderness with all her being. This outcome is quite distinct, we argue, from being wronged, assaulted, raped, or murdered (the originary traumas of many South Indian goddesses), since her movement is inward, into selfness.

The effect of the transformation is an even closer tie between the goddess and kingship, the tie of sister-goddess and brother-king. The goddess apparently does not give birth to kingship, but she does nurture it. Yet Paidimamba insists that on her return she be worshiped across the lake from the city, on the other side, distant from the civilization that she has returned to nurture and protect. So in her return there is a gap, perhaps a rupture that will have to be mended or sealed, in order for her to fully emerge and develop as this protectress. This gap or distancing may be integral to the rhythms of cosmic surface, of decline and re-growth; it leads annually to her rebirth and maturation. It is here that human beings enter, helping to nurture the Goddess into presence within civilization, so that she will foster the human—that is, the kingly—portion of her cosmos. Her rebirth is the coming into infancy of some aspect of hers that must be shaped, grown into presence, in order for her to perform her own nurturing role. The holism of her cosmos is not a given—it is not a static nor a homeostatic whole, one automatically renewed through mechanistic ritual. Paiditalli's cosmos must be practiced into existence (with all the emergent uncertainties this entails) in order for its evolution to become auto-poetic, that is, self-organizing. As she grows into presence within the human portion of her cosmos, she comes closer to human beings and their world, acting powerfully within it. No birth, not even Paiditalli's birthing of herself, is certain as the cosmos seeks its own dynamics of origination within its own, often murky, uterine depths.

As Bairagi Nayudu tells the story of Paiditalli's origins, she and her brother had neither mother nor father. They grew up under the guidance of Appala Nayudu. He was like a foster-parent to Paidimamba. After Paidimamba jumped into Big Lake, Appala Nayudu searched for her. 'They looked around and saw bubbles coming up from Big Lake. As soon as they saw the bubbles, someone jumped in and pulled her out. She was at her last breath [*kona upirito undi*]. "What has happened to my brother?" she asked. In such a situation, you can't tell the truth. Appala Nayudu lied, "He is alive." She refused to believe him. As soon as she said that, he burst into tears. Then she understood the truth. She died, breathing her last in Appala Nayudu's lap [*odilone*] ...'[20]

Appala Nayudu became sick at heart. He was now near the Wilderness Temple; he stayed under the neem tree there. When people told him to

20 It is also said that Paidimamba 'left her breath.' An English speaker would commonly say that 'her breath left her'. Might there be implications, perhaps yogic, that in leaving her breath, Paidimamba separated from herself, from her self, rather than suicidally killing herself?

go home, he said, "I want no *aihika-sambandham*, no connection to the affairs of day-to-day life. I'll stay right here." That night the girl appeared in his dream and said, "Appala Nayudugaru,[21] I have become Ammavaru [the Goddess]. The image relating to me [*na taluka vigraham*] is in the lake. Bring it out and perform puja to it." Next morning they looked for the image, but in the first attempt they could not find it. They sought the help of the Jalaris, who looked for it and found it. Appala Nayudu prayed to Vijayavada Kanaka Durga: "What have you done to this devotee of yours? You made my stomach ache. From the lap of her mother she came on to my lap. That's why I can't bear this suffering." Kanaka Durga said, "Establish the image of this girl and I will bless her from the top of the *dhvaja-stambha*, the flagpole. That flagpole is the Sirimanu."

In Bairagi Nayudu's version, brother and sister, orphaned, are aided and succored by his ancestor, Appala Nayudu. She dies in his lap and returns to him as the Goddess, coming in a dream, telling him to find her image and install it. Appalu Nayudu is something between mother and midwife to the young Paidimamba, as is clear from the role of the lap in Bairagi Nayudu's story. Appala Nayudu holds her on his lap. She dies in his lap. He complains to Kanaka Durga, 'You made my stomach ache (*kadupu sokam pettavu*). From the lap of her mother she came on to my lap.' The lap is commonly associated with womb (the same word can serve for both); the girl 'swings' from the lap of her mother to that of Appala Nayudu.[22] In Bairagi Nayudu's version, Appala Nayudu is also the first to be carried by Paiditalli in the Sirimanu. So, too, in the (re)birth of Paiditalli, discussed below, Bairagi Nayudu has something of the role of midwife; and as he is carried aloft by Paiditalli in the Sirimanu, he is the priest who enables in a 'motherly' way the emergence of king and brother—Paiditalli brings him to his home, to the Kota palace-fort that is the heart of the kingdom.

So far our discussion of the Vizianagaram cosmos implies that there is a dimension of intimacy in the goddess's relations with human beings. Her sense of intimacy and caring is an interior quality that relates to the depths of others or, more accurately, to the deepening of these depths. The well-being of human beings matters (perhaps more than before) to the transformed female. Paidimamba's intimacy with the goddess is perhaps most evident in her self-sacrificial rebirth within the depths of the goddess,

[21] *-garu* is a Telugu honorific suffix.

[22] The swing plays an important role in closing out the presence of Paiditalli, two weeks after the Sirimanu.

and her self-intimacy will be emphasized further through her annual (re)birth. Most telling, this deity is female, endowed with an intense sense of intimacy axiomatically denied to most males, who traverse and build atop shallow surfaces and flat lands.[23] Qualities of intimacy need to be highlighted here in order to understand, further on, the annual (re)birth of Paiditalli and her evolving relationships with Vizianagaram.

3. *Historical Excursus*

We know something of the foundation and rise of the Vizianagaram state in the first half of the eighteenth century (although the Pusapati dynasty traces its origins back to one Madhava Varma in the seventeenth century, from Pusapadu, near Vinukonda in the southern delta).[24] Devotees of the golden goddess Kanaka Durga of Vijayawada, the Pusapati kings set off—allegedly with the blessing of the distant Mughal dynasty in Delhi and its nominal subordinates in Hyderabad—to seek their fortune in the wilderness (*manyam* or *manne*) of southern Kalinga, that is, the northern Andhra coast.[25] Officially, the Pusapatis became tax-farmers within the Hyderabadi-Mughal system and consequently won the high-flown title of *Manne Sultan Bahadar*, "Sultans of the Wilderness."[26] In reality, they were largely independent competitors in the century-long tug-of-war among a host of local *pallegallu* rulers for control over the region's rich resources. By the early years of the eighteenth century, this struggle had eliminated all but a few major players: the Pusapatis, ruling from Kumili; a tribal kingdom centered on Madugula, but with its ritual center situated at Paderu

[23] This generalization does not include the depth specialists within civilization—playwrights, poets, Tantric yogis, Ayurvedic healers and, occasionally, perhaps kings. All healing involves the opening up of depth in order to repair—whether human beings, kingdom, or cosmos. There seems to be hardly any way of healing on the surface without discovering interiority, the hidden, the vital, the speedy and the accelerating. Healing is done from within, not from without. Consider, too, the poet as depth-specialist, creating through the involution of language; think of the interiority of yoga; note the bodily entanglements of wrestlers. On the cosmology of wrestlers in North India, see Alter (1992).

[24] See *Pusapati rajulayokka purvottaram*, 1; discussed by Berkemer 1993: 252. Our thanks to Velcheru Narayana Rao for reading the Vizianagaram *kaifiyyat* with us and commenting upon it.

[25] According to the *Ranga-raya-caritamu* of Dittakavi Narayanakavi (1.115), Madhava Varma came to this region together with a Mughal adventurer called Sher Muhammad Khan; for his services, Madhava Varma was rewarded with the Vizianagaram kingdom. See Narayana Rao, Shulman, and Subrahmanyam 2001: 63.

[26] *Pusapati rajulayokka purvottaram*, 4.

in the high hills of the Eastern Ghats; and the low-land fortified city-states of Sringavarapu Kota, Golakonda, and Bobbili, mentioned above with reference to the origin myth of Paiditalli. Our sources, mostly kaifiyyats written by village accountants (*karnams*) at the request of the famous Colonel Mackenzie in the early nineteenth century, allow us to reconstruct the final stages of this regional conflict; largely through the ruthless machinations of a wily Diwan, Burra Buccanna, a paradigmatic figure for this newly crystallizing polity, Vizianagaram succeeded in dominating its rivals and drawing them into an emergent fiscal and military system operating from the new capital of Vizianagaram—founded in the second decade of the eighteenth century, according to tradition on Vijaya Dasami (Victory Day) in the autumn Navaratri festival celebrating Durga as Mahisasuramardini, her killing of the great Buffalo Demon, in the Vijaya (Victory) year 1713–1714.

Several traditions speak of the founding of this aptly named "City of Victory" (more properly transliterated Vijayanagaram; we have retained the Anglo-Indian spelling in order to differentiate it from the well-known imperial state far to the south in the western Deccan, at Hampi). The site was, not surprisingly, chosen for security considerations; large hills protected it to the east and north; to the west was the village of Devulapalli; the fort-palace was constructed on the edge of a lake fed by mountain rivulets.[27] Originally, there was nothing here but wilderness and a small village called Rega; the area was full of tigers and subject to predatory raids by Muslims (*mlecchulu*). Still, after the founding of the fort by the first truly regal figure, Peda Vijaya Rama Raju, the entire Pusapati clan, with its women and children, marched there from Kumili in four days.[28] Some say that Peda Vijaya Rama Raju first discovered the site on a hunting expedition; to his amazement, his hunting dogs were savagely attacked there by the rabbits they were hunting. The king asked a Muslim ascetic, Denkha Shah Wali Baba, who was meditating in the shade of a tree, about the meaning of this strange reversal. Denkha Shah said: "Kumili will be ruined. Build a new fort here. Heroic manliness (*paurusham*) is in the soil." So Vijaya Rama Raju moved his kingdom to Vizianagaram and built the first, mud-walled fort.[29] An oral account collected by Georg Berkemer

[27] Ibid., 9–10.

[28] Ibid.

[29] The present day fort is 'a great square erection of brick and stone measuring about 250 yards each way, surrounded by the remains of a ditch, and having a big bastion at each corner' (Francis 1907: 336).

insists that the founders of the fort—two royal brothers[30]—deliberately ignored the Muslim saint's directions for building the fort nearby, instead constructing it precisely on the spot where the hunting dogs were attacked by a (single) hare; as a result, the new state lost its chance to conquer the entire world.[31] Denkha Shah's tomb is situated today about halfway between the royal fort and the Square Temple (*saduru gudi*) of Paiditalli.

This myth of origins fits a pattern attested from all over south India in this period; but we need to emphasize again the self-perception of this polity as a wilderness kingdom, only one step removed from tribal polities such as Madugula and Jeypore,[32] remote from the more settled and stable political systems of the delta and the interior and permeated by the unpredictable powers of the undefined. This is a kingdom of tigers and tiger-like warriors whose peripheral domain is alive with sorcery, magical Yoga of a distinctive character, tribal deities, fishermen, hunters, and goddesses at home in the wilderness. Conspicuous by their absence are the Brahmin communities usually associated with patterns of political homeostasis in south India. Although we do hear, occasionally, of royal land-grants to Brahmins,[33] and we can also trace the rise of large-scale Sanskritic temples (at Mukhalingam, Rama-tirthalu, Srikurmam, Padmanabham, Palukonda, and so on) that were drawn into the orbit of Vizianagaram patronage and control in the course of the eighteenth century, it seems that the Vizianagaram state managed quite well without Brahmin authority or assistance. This state rested on quite distinct forms of legitimation and generative power, the kind of power that derives from alchemists, itinerant magicians, musicians, poets, pandits and grammarians, wrestlers, tiger-dangers, and unconventional Tantric Yogis. But it was Paiditalli, above all, who provided, in a sense, a practical experiment for the preservation and enhancement of the political domain, including the annual renewal of kingdom and kingship as a going enterprise. We return to the character of this 'experiment' in the conclusion.

As we have already seen, a major crisis overtook Vizianagaram in 1757 with Charles Bussy's raid into the Northern Circars and the Bobbili war, ending with the destruction of the Bobbili fort and the consequent

30 This seems to be a memory of the historic rivalry between the brothers Peda Sitaramaraju and Peda Vijaya Rama Raju.

31 Berkemer 1993: 275.

32 See Schnepel 1996 for a study of this kingdom situated to the north of Vizianagaram but linked in many ways to the historical evolution of Pusapati kingship.

33 Thus an early king, Raghunatha, is said to have given two or three agraharams—tax-free lands—to Brahmins: Pusapati rajulayokka purvottaram, 3.

assassination of Vijaya Rama Raju, the Vizianagaram king, by the Bobbili hero Tandra Papayya (whose immense sword is still on display in the Bobbili palace). Somewhat surprisingly, this traumatic event is "celebrated," one might say, in Paiditalli's autumnal Sirimanu procession: out of the brother's tragic death comes the sister's full emergence into divinity. But the initial tragedy at Bobbili was only the opening scene in the drama of Vizianagaram's imperial decline (perhaps self-immolation would be a better term). Vijaya Rama Raju's successor, Ananda Gajapati, abandoned the French connection and prudently threw in his lot with the British, who indeed gained full control over all of northern coastal Andhra by 1765. Over the next decades, the Vizianagaram kingdom maneuvered rather desperately, with some initial successes, in the hopes of surviving the remorseless expansion of British power and control. By 1794, the last fully independent ruler of Vizianagaram, yet another (China) Viyaya Rama Raju, had had enough; faced with the British demands for ever greater payment of revenue/tribute, he decided to make a heroic last stand on the field of Padmanabham, not far from the capital. He was, predictably, cut down in a suicidal battle with the English forces, under the command of one Lieutenant-Colonel Prendergast, on July 10, 1794.[34]

Vizianagaram's final loss of military power and absorption into the British system was not, however, the end of its creative political role. It is something of a cliché in modern historiographies that such little kingdoms compensated for their reduction to subservience by hypertrophied investment in the cultural domain, and in rituals of display (Dirks 1984; Waghorne 1994)—as if such rituals were not, in any case, the very substance of south Indian kingship and the stuff of power. In the case of Vizianagaram, we would perhaps do better to speak of a natural intensification of the inward-directed existential experimentation proper to life on the cusp of the wilderness. The exercise of brute military force and the play of imperial politics were probably never Paiditalli's primary concerns. Her extrusion of interiority to overcome loss in the surfaces of civilization probably was. Nineteenth-century Vizianagaram transformed itself into the most vibrant cultural center between Calcutta and Madras, a site of

[34] On the battle of Padmanabham as recorded in Chatrati Lakshmi-narsa-kavi's *Padmanabha-yuddhamu*, see Narayana Rao, Shulman and Subrahmanyam 2001: 79–92. For the turbulent power struggles of the 1760's, 1770's and 1780's at Vizianagaram as masterminded by the Diwan Sitarama Raju—a man fashioned in the model of his predecessor, Burra Buccanna, mentioned above—see Sundaram 1946. See also Sarma 1994 for an overview of Vizianagaram history.

truly astonishing creativity and, ultimately, the arena where what might be called a fully modern awareness first came to be articulated in Telugu. This highly specific version of "modernity," associated with the names of the beloved ruler Ananda Gajapati (1850–1907) and the poet-dramatist Gurajada Apparao (1862–1915) (Rama Rao 1985), emerged organically out of the peculiar cultural-political configuration of the northern Andhra coast, whose somewhat volatile components we have already listed, including a strong notion of introspective self-cultivation in the Kalinga style of Tantric Yoga.

Vizianagaram remained throughout the nineteenth century, indeed remains still today, a wilderness kingdom, held together by a wilderness goddess slowly ripening and fermenting for half of each year in her shrine proximate to the royal palace—a goddess brought every year into the heart of the kingdom and then sent back to the depths just beyond it. The modern history of this kingdom is thus not only, or even primarily, the record of military defeat, dynastic decline, and the slow refashioning of collective selfhood in a new mode, within the terms of a transformed political system.[35] It is at least as much the history of a goddess intimately known, indeed nurtured and grown, year after year, through ritual and aesthetic means, within the institutional and spatial core of the state. Put differently, modern Vizianagaram history is largely the history of the ongoing inner life of this goddess, her moods and changing conditions, her somewhat unpredictable states of mind, her internal shifts and necessary periods of disequilibrium, her evolving awareness, and her unfolding imagination of herself in relation to her subjects and devotees.

4. *Goddesses are Grown (Gods are Not)*

We have stressed the notion of intimacy as integral to Paiditalli. Intimacy with goddesses is highly layered, intensified, and textured within the home, and in order to foreground this notion we turn to another venue in Andhra Pradesh—in the near past and not far from Vizianagaram. Many homes in the villages of the Godavari Delta, where M. V. Krishnayya grew

---

[35] Indeed, it is our strong impression that Vizianagaram kingship today is recognizably continuous with its roots in the eighteenth and nineteenth centuries, despite the formal abolition of the royal role in independent India, the royal family's successful transition to democratic state politics, and the death of the last king, the beloved P. V. G. Raju V, in 1995. See further discussion below in section 8.

up, have a goddess living in the kitchen. Her name is *Taravani,* a certain kind of rice-gruel.[36] Raw white rice is always washed before cooking. If the water used for washing rice is poured into a large pot, sitting in the auspicious north-east corner of the kitchen, this fluid will begin to ferment into the presence of the goddess. Arranging the location, establishing the conditions, beginning the process of fermentation—all these are the results of human action. In this sense *Taravani* is *enabled to grow into presence* in the north-east corner, across from the hearth and its fire in the south-east corner.

Yet the goddess who is grown is not a construct of the human. The human provides an entry point—the vessel, its fluid contents, its fluid depths—into which the goddess flows, configures, takes shape, intensifies. The vessel, the pot, is then something of a womb *of* the goddess, within which she grows herself. Human beings provide the conditions for growth, but the coming into presence of the goddess occurs wholly on her own terms, as do her effects within the home. There is no way of forcing *Taravani* into a (family) kitchen where she does not belong, without inviting trouble.

The north-east is an auspicious direction, the right place for the well in the courtyard, from which water is raised from the depths. The *Taravani* pot sits in the north-east corner of the kitchen on a *kuduru,* a low, round, ring-like stand made of woven palmyra leaves. The form of the pot, *kunda,* is that of completeness curving into itself, without angles, without corners, without rupture. The word *kuduru* means 'grounding', 'settling', 'matching'—the goddess settled in a stable, harmonious mode of being suited to the particular family. Put otherwise, *Taravani* is made to feel at home within the female domain of the kitchen, living there in the proximity of family members. The north-east corner of the kitchen is a shrine to *Taravani*: the *kuduru* stands on flattened earth covered with white floral designs (*muggulu*) of lime, intended to entice her in. On the wall above the pot the face of the goddess is painted in turmeric (*pasupu*), a circle divided into three parallel segments, with a dot (*bottu*) of vermilion (*kunkum*) in each segment.

The essences of the goddess, indeed of all females, are said to be *pasupu* and *kunkum,* respectively of golden and red color. Gold is thought to be the solid closest to liquid. Turmeric, though applied to the exterior of the female, to her skin, enlivens her interior life, *layering* her from within,

[36] Taravani is identified with Lakshmi, goddess of bounty and prosperity.

adding depth and texture. She glows more intensely from within, a self-shining burning with the brightness of life-force, itself focused by the dot of vermilion on the forehead. Her self-shining seems also to signal the *intensification* of coherence and integrity, the qualities of selfness, the qualities of depth. Turmeric is thought to open space for the creation of life within the female. In our terms, turmeric creates depth, opening space for the vitality and beauty it generates from within. Turmeric and vermilion open, focus, and intensify space and depth for the goddess in the human world. She enters into this domain, becoming (gently) present in the human world, comfortably at home here. Thus the goddess grows from within herself into the space and depth opened up by turmeric and vermilion. Turmeric enables the *self-layering* of the interior of the goddess, thus rendering her interior more complex. Turmeric is a crucial agent in the growth and development of the goddess in the human world, indeed of the female generally. For a mature woman not to have turmeric is inauspicious and dangerous. Vermilion, on the other hand, gives to a woman the honed feeling that she is specifically whoever she is. The *bottu* dot, directing itself inward, is a focus for female interiority; directing itself outward, it absorbs and nullifies evil aimed at the female from the glances and thoughts of others—these are drawn towards the vermilion dot. A woman without a *bottu* lacks focus, essence, significance.[37]

Through the essences of turmeric and vermilion, goddess and woman overlap, periodically, momentarily. Every morning, after bathing, each married woman in the house smears her face with a light coat of turmeric paste and arranges the vermilion dot, thus participating in the being of the goddess. All married women share this condition and form.[38] Again, turmeric and vermilion seem to add *layers* to the core being of the female, and this *layering*, to which we refer below, indexes the ongoing maturation of these female qualities. *Pasupu* and *kunkum intensify* these qualities, attracting them, drawing them from the interior space of the female to her surface, to the interface of her interaction with others.[39] Layering and intensification are dynamic properties of goddesses and married

[37] Marriott (1998: 299) comments that, in India, females have 'greater fluidity and internality' than males, and thus, 'greater inner-bodily space'. By contrast, males are of 'putatively harder, cooler, less open natures.'

[38] Unmarried girls and widows do not do this. The essences of the female have not yet awakened in the unmarried girl and are to be extinguished in the widow, who in many ways becomes one of the living dead.

[39] The Telugu term, pasupp'ekkincadam, 'applying pasupu', could also be translated as 'intensification.'

women. This is why they, and not unmarried girls, apply turmeric and vermilion. For married women the intensification of these qualities has its own temporal rhythm: these qualities come in the morning and wear out by late evening, to be re-created or re-born the next morning. At night it is important for the housewife to revert to her entirely human nature; she sleeps beside her husband not as a goddess but as a woman. A husband would be terrified to find a turmeric-bedecked goddess in his bed. A married woman is a kind of sequential divine-human combination: a goddess from early morning, a human being from late evening onwards.

Within the uterine pot, the fermentation of the goddess is understood as her self-distillation, her self-intensification, her presence in the home becoming more and more vibrant, lucid, focused, and, not least, fully female. These dynamics are spoken of explicitly as *layering*, which is also of direct relevance to the (re)birth of Paiditalli. We can describe this process as her coming into presence from deep within herself toward her interface with the human world in order to emerge here within womb-like depths, in the kitchen, the domain of women. To develop through these trajectories, the Goddess needs the proper conditions for interaction with her immediate environment. When these are satisfactory, she self-develops, self-intensifies, emerging within herself within the household as wholly herself, with refined capacities for helping members of the household. This dynamic of *intensification*—of coming into being as wholly herself, and thus as fully harmonic with the cosmos that she is[40]—points to the fullness of her cosmos and, indeed, of herself. Intensification increases intimacy with herself—as self-feeling, self-knowing—and with others as they take her (her cosmic harmonics) into themselves, or into their selves. What, after all, is more intimate than this movement?

The presence of *Taravani* in the kitchen cannot be taken for granted. Her vibrating presence requires continuous concern for any change in her condition of becoming. If attention lags and turbulence ensues, the goddess will be gone. She can be broken, her continuity of self disrupted, her equilibrium disturbed. The housewife must consistently calibrate the

[40] South Indian deities must be persuaded and cajoled to emerge from their depths, to intensify in the human world (see, for example, Chapter Three in this volume). This happens on a daily basis in temples, as the deity is awakened, washed, dressed, shown to herself in a mirror, fed, and thus made fully present to receive worshipers. Joyce Flueckiger (n.d.) suggests that covering the visage of the goddess, Gangamma, in Tirupati, with pasupu is integral to a process by which the goddess recognizes her true self (visvarupam). One could add that when the goddess *layers herself*, she then recognizes the fullness of her own deep interiority, without limits, that is her true self.

intensities of the goddess, scrupulously making necessary adjustments. Moreover, the relationship of feeding is reciprocal. Whenever food is cooked, a part is presented at the base of the pot, a gift to the goddess, while the rest is eaten as her *prasad*, her gift of bounty to the people of the household. Within this local yet most pivotal of domains, the kitchen is a micro-temple, the housewife something of a priestess.

Therefore, great care is taken to observe the needs of *Taravani*. Like all goddesses, she is not to be rudely disturbed, and great sensitivity is shown in all contact with her. On Tuesdays and Fridays, 'goddess days' in Andhra, she is not to be touched, for she is deeply within herself in a condition of non-activity and self-absorption, distant from the human. But if at any time she is touched inappropriately, abruptly, she becomes angry, and the fermenting water spoils, becoming sticky, semi-solid, with a bad smell. Then the liquid must be thrown out, the pot washed, and everything re-arranged in the ritually correct ways. The mother of the family is the only person who handles Taravani, cleaning her pot, reworking the earthen floor with cow dung and *muggulu* florals, repainting the goddess on the wall. Before attending daily to Taravani, the mother of the family must bathe, as she would before entering a *puja* room or before performing a *puja* in a shrine. The profound closeness of deity and human is crucial in this organic, densely intra-related cosmos. Deity and human are different from one another, yet not in any abrupt or absolutist way; their presence, indeed their co-presence, is mutually and densely intra-woven.

Within the intimacy of family, the deity is present to help (especially to protect children from disease). As she ferments, as she grows into presence within the fullness of the uterine pot, she is consumed, drunk, often daily, by all members of the household (and especially in times of disease by neighbors of all castes). The drink gives *calladanam*, coolness, cooling the body's system like a purgative. The fluid entering one's inside is not a byproduct of growing the goddess—it is the goddess herself as sustenance entering one's intimate interior, spreading throughout one's body-being. Nurtured, interacted with, matured, distilled, tasted, internalized, digested, the goddess is in intense relationship with the home. Once she has been ingested the goddess protects and heals. Consider that drinking and eating are among the most intimate of human actions. They take place at the interface between the interior of person and the exterior world. One must take the exterior world into oneself in order to drink, eat, survive, in the most mundane and taken-for-granted way. Yet there is always potential risk in this, even with the most zealous and careful of preparations. There is always in eating and drinking a powerful element of trusting the exterior. This is probably no less true for the Goddess who exposes herself to the ministrations of human beings as she intensifies into presence.

The idea of inner coolness, *calla*, located in the stomach of the human being, is widespread in the Telugu-speaking world. *Calla* implies anything cool, comfortable, secure, free from anxiety—a matter of existential well-being, the most sought-after state. Inside one's stomach the presence of the goddess feels good, a continuous, heavy force that is the pleasure of ongoing anticipation, a harmonics of gravity, holding one intimately to the earth, connecting, rooting in this center. Yet *calla* is not always perceptible or accessible. A goddess may be required to activate it or to allow for awareness of its presence, and so to have contact with it; and there is anxiety lest this inner, foundational *calla* be lost. External heat can almost eradicate it. Human interaction threatens it. Food affects it—hence the concern to eat cooling foods. A person from the outside can dislocate, disembowel, diminish or extract one's interior coolness, indeed one's inner cool. And the very existence of the interior *calla* may be the necessary condition for prayer, meditation, healing. The goddess amplifies the inner *calla* and insures against its being wasted or lost, thereby enhancing the innate immunity of the human being to harmful intrusion. The process of fermentation in the *Taravani* pot is itself a *calla* condition, intensifying the goddess's own inner dynamics of being in the human world, of becoming present to human beings. Fermenting is perceived as distilling, as refining and condensing the being of the goddess, enabling heightened contact with her. By distilling the goddess, human beings go beyond their narrowness, weakness, and fragility. They impact upon and shape *her* life, and therefore their own.[41]

As we return to the rites of Paiditalli, consider once more the profound fluidity in this Telugu cosmos. The dynamic of fluidity is that of continuous movement—in other words, of fluid intensifying itself, continuously flowing into and through itself in ever-shifting eddies and currents,

[41] When the local goddess (whoever she may be) comes to visit a neighborhood, lane by lane, house by house, she brings coolness, *calladanam*, in the form of big pots carried on the head. She is offered Taravani fluid. The two goddesses, two distinct authorities with differential responsibilities, happily join together and enjoy a certain companionship through the exchange of Taravani. They enrich each other, recharge one another at different levels. The active fermentation at the right consistency is a critical medium for this kind of interaction between deities, enabling them to make contact with one another through one another, without going outside each other. A certain sense of elation accompanies this meeting—also in the spectators, the householders who are paying close attention to what happens at such moments.

perhaps fermenting and self-distilling. Water inherently has depth in this cosmos (even when it appears shallow) because it must flow through itself in order to continue to be the fluid it is. This water inherently intensifies itself. The fluid totally fills whatever space it flows through. Consider now that Paidimamba plunges herself into the fluid domain of Big Lake—and drowns. She does this in order to become one with Durga. Implied in this sequence is a vision of the Lake as a medium of wilderness, an entry into the self-intensifying fluidity of the goddess within herself, that is, into her deep coolness. Big Lake is something of a continuous container within the topology of the human world. Paidimamba's choice of death is eloquent and profound: she enters the healing depths of the Goddess, 'losing' her self there. Perhaps Paidimamba's desire is indeed to *intensify* herself through Durga by joining the goddess through self-sacrifice. Death by drowning is the taking in of fluid until one is filled utterly (and can no longer breathe, 'leaving one's breath'). The cool waters of depth, of Durga, fill the interior of Paidimamba, and she joins the Goddess, becomes intimate with the Goddess, becomes the never-ending movement of the Goddess in the ooze of the lake bed. In the process the Goddess too is changed, becoming closer to, more intimate with, perhaps even more beneficial to king and kingdom. Though only implied, the logic of this configuration seems quite similar to that of Taravani in the houses of the Godavari Delta.[42]

Yet there is a cosmic caveat to the interior fluidity of Paiditalli, or of Big Lake as the topological interiority of the Goddess. Once she leaves the Lake, leaving the wholeness of her interiority within herself—within the

[42] In contrast with female deities in this area, male deities have difficulty with what we are calling depth and its creative, healing potentials. Consider the greatest of these male deities, Simhadri Appanna, the Varaha-Narasimha form of Vishnu, whose mountain temple of Simhachalam was included in the territory of the kingdom of Vizianagaram for some two centuries. The kings of Vizianagaram used Simhachalam as a retreat whenever the British forced them to leave the city. In one prominent myth, a hunter came upon the god in his Varaha (boar) *avatara* form rooting in the forest of the mountainside. The hunter shot a poisoned arrow into the boar, only later realizing that this was God he had wounded. The hunter, like many other people of the depths, was also a healer, and he offered to cure the god's wound by applying balm—but he also noted that a scar would remain. The god angrily rejected the offer: "What use is healing that leaves a scar?" In the *garbha-griha* of Simhacalam, Simhadri Appana stands, his feet stuck in the earth, unable to move, his head and body covered entirely by an oval of sandal paste said to salve his wound. Tradition tells us that the wound has been infected from the day the hunter penetrated the god, and that it suppurates, so that the priests must regularly drain it of pus. Only once a year is the sandal paste removed from the god. On the one day a year when he comes out of his temple and rides his chariot around his cosmos (the temple), the chariot is unable to begin moving unless fishers, again people of the depths, pull it.

greater cosmos that is her self—she begins to acquire fixity of form and the desire of intentionality within the human world. This fixity of the Goddess is initially related to the relative solidity of earth and to the rootedness of natural form in earth—and then, as we will see, to the uprooting of natural form from earth, to its acquiring motility in the human world, even as this fixity of form moves again toward distance from the human, as the Goddess returns to the greater fluidity of her own deep interior. Paiditalli's cosmic career as she surfaces from Big Lake joins water to earth through her changing being. And, though Paidimamba entered Big Lake with the declared intention of re-turning and re-surfacing, the annual transformation of Paiditalli from the fluidity of interiority to the relative fixity of form on the flatness of the human world cannot be taken for granted. The transform is not mechanistic; rather, it is a moment of high ritual uncertainty, since Paiditalli is at times resistant to emerging from within herself. Her coming should be understood as a cosmic lurch through discontinuity, perhaps a self-sacrifice of some part of her fluid selfness.

5. *The (Re)Birth of Paiditalli: The Ritual of* Devara Pandaga[43]

The year of Paiditalli is composed of two periods, roughly six months each. The goddess is re-born in the Telugu month of Vaisakha (April–May), at the height of the dry season, yet towards its end, in the depths of Big Lake close to the *vanamgudi*, the Wilderness Temple. As she enters the human world, she acquires form that is quite formless; and initially, perhaps, she has little intentionality. Months later, as her active mode reaches its highest intensity during the Sirimanu Jatra in the month of Asviyuja, her form is sharply delineated, aimed with direction, force, momentum. During the intervening period she is given optimal conditions to grow her form, to form herself, to intensify her presence, to hone her intention. Two weeks after the Jatra she returns to the *vanamgudi* for six months. Within wilderness and its depths, understood in manifold ways, her presence fully enters her self, once again distant from the flat surfaces of the human world, undisturbed, uninvolved in human affairs, until she is asked, or pleaded with, to come forth once more.

In the month of Vaisakha, as agriculturalists get ready to plant the first seeds of rice, the staple food crop of the region, the coming of the great

[43] Our observations of the Devara Pandaga rite were carried out in May 2004 and May 2006.

south-west monsoon rains is essential. Paiditalli will bring the rains, and much more besides, for the growth and fruitfulness of kingdom and kingship ... if she appears. Paiditalli's coming is an event of creation in its own right, independent of anything that has gone before. Such events are self-generating, dependent on their own cosmic impetus; thus the goddess must desire to emerge from within herself into the flat world of human beings. We have likened Big Lake to the womb of the Goddess within which, or through which, she gives birth to herself, perhaps through a sort of parthenogenesis in which her infant and mature selfness will exist simultaneously, overlapping, the former poised on the lip of the human world, the latter distant and deep within.[44] Since she is cosmos, the whole, she is simultaneously all of herself and any portion of herself, the whole curving within the whole. Nonetheless, that infant portion of herself that she will entrust to human beings is vulnerable, and so there is potential trauma, or indeed actual trauma, in her birthing—though the form(s) she develops during the months after her self-birth will gain strength, power, and self-assertion and are akin to the evolution and maturation of cosmos, or at least that portion of cosmos in which human beings live.

These perceptions may be somewhat distant from those of her priests and worshipers. People know and feel the hot season as a time of desolation, the heat searing vegetation, burning life; a time of destruction, of disease (more so in the past), of disease-bearing goddesses, of death, especially the death of the vulnerable young that is the death of future. A flat world of hard, dry surfaces. A time of waiting, of hoping, for the resurgence of life. The farmers know well that nothing is guaranteed, certainly not the meeting of rain and earth in fruitful union. In a way, the hot time is an end-time, a period during which Paiditalli is absent from the human world, deep within the potential fruitfulness of wilderness that is most separated during this time from the human world. Yet this period of destruction also seems to clear the human world of detritus and debris. If this end-time is to pass, then the rains must come to melt the flatness of earth, to open the potential for growth that only exists within depth. For the rains to come, Paiditalli must come. For Paiditalli to come, she must be drawn out of herself into the human world, where she will be born out

[44] To explicate this: Paiditalli, her cosmos, overlaps with herself. Paiditalli does not disappear from the Vanamgudi, but continues there deep within herself. So there is no end to cosmos, no rupture and then a new beginning. Rather, there is an attenuation of the interior fluidity of cosmos as an aspect or qualities of this rise to Paiditalli's own surface, acquiring forms, while she, her ultimate true form, stays fluid, deep within.

of herself, and where she will become increasingly present, increasingly vital, more and more herself like a pregnant woman (though always a virgin) ready to give birth to life. Such birth-from-depth is a dynamic through which life emerges from within itself, until its impulse for emergence is momentarily suspended (however lengthy that moment may be).

Remember that, first and foremost, Ammavaru is female in a social order in which the birth and nurturing of offspring may be the greatest personal joy a woman will have during her lifetime. When we come to discuss the growing of rice, it will become evident that this dynamic of emergence from the depth, that is, of growth as pregnancy, birth, and nurturance is understood in precisely these terms by farmers of this region.

Like ritual complexes generally in India, that of Paiditalli is activated by positioning its elements so that they interact, generate, and self-organize their potentialities. Persuading Paiditalli to give birth to herself in the human world is not quite a straightforward ritual matter. Though she cannot be coerced through ritual, she can be drawn into the nexus of family, into warmth and intimacy, and shown how much, how deeply, she is respected, loved, appreciated. In this way one may awaken her to the needs of her human worshipers and their world. She has, of course, a particular affinity with women, and it is the women of Bairagi Nayudu's household who open the way to her coming by visiting her in the *Vanamgudi*, the Wilderness Temple, where she is said to be resting. This visit occurs one week before the *Devara pandaga* ritual at Big Lake.[45]

The day before their visit, they and the men visit Paiditalli's original shrine within the *uru*, tucked into the three-species tree in back of the *Cadurugudi*, the Square Temple. Inside the tiny shrine are a series of almost aniconic wooden carvings, the *Peḷḷu*, each with a crude face, a new one added every year, cut from the Tamarind tree that was Paiditalli in the previous year's Sirimanu Jatra. The Peḷḷu (who are both Paiditalli and her younger brother, Potu Raju) have a crucial role in the Tolellu rite on the evening before the Sirimanu. Family members smear *pasupu* on the figures within the tiny shrine, decorating them with vermilion *bottus*, stirring the Goddess of past Sirimanu Jatras, at the apex of her evolution in the human world, to meet the Goddess who they hope will emerge in the near future, newly born, from Big Lake. One may surmise that they also stir the deep connectivity between older sister and younger brother. Then they go to the palace-fort, the Kota, to visit the mysterious Kota-Sakti

[45] In 2006 their visit took place on May 8th.

(who seems never to leave the fort and who is the continuous protector of kingship there) and to perform a puja to her, inviting her to be present when newly-born Paiditalli arrives. Paiditalli, the goddess of kingdom and kingship (*illavelupu*) and the Kota-Sakti of the royal palace have their special closeness, and guests should be invited personally, as to a wedding. Printed announcements (*catimpu*) of the imminent arrival of the Goddess are distributed; the various functionaries and those who will wear guises (*veshalu*) in the processing of the Goddess are informed of the coming event, as are the households of the neighborhood lanes where Bairagu Nayudu lives and through which the procession will move.

When the announcement of the intention to bring Ammavaru is made, the uru—the original city built on the plan of a mandala—is tied (*uru kaṭṭaḍam*) or bound. This is done by a member of the oil-presser caste (Talaiyari). Water is mixed with turmeric and the oil-presser spills this as he walks the boundary of the uru, marking and sealing this.[46] This binding of the boundary can be understood as the activation of the original mandala for the period of the goddess' presence there. This is done, people say, so that Ammavaru will feel cool (*calladanam*) and comfortable; so that she will not suffer any potential threat, for she is very vulnerable while in the uru, especially so when she is an infant.

The day after visiting the tiny shrine in back of the Cadurugudi, Nayudu's wife, his eldest daughter, Nirmala, his youngest daughter, and three young unmarried girls (one the daughter of Nirmala) pile into an auto-rickshaw on their way to the Wilderness Temple. Bairagi Nayudu's youngest son, Venkat, follows on his scooter. Some days earlier Nirmala had said that they would go to Ammavaru to tell her that shortly she would 'have to move house'. Indeed, she has her own room, her own home, in the Wilderness Temple. But as this visit develops, this imperative, or even the declarative, will be absent. All the women are beautifully dressed in saris and jewelry, with *pasupu* on their faces. With them are steel vessels full of cooked food for the Goddess. In 2006, Nayudu himself is too ill to participate. They arrive at the Vanamgudi at around 1 P.M. There are few people about, apart from four priests who clear the entire temple space for the visitors.

---

[46] In past times in villages in the area, once the boundary was sealed no marriages took place during this period; there was almost no interaction with the outside world; and if someone did have to leave, this person would first offer turmeric and vermillion to Paiditalli.

The women perform a puja to Ammavaru, familiarly playing with the her *murti*, decorating her, touching her the way a bride is touched by her female kin. The mood is happy, even joyous. They offer Ammavaru *pasupu* and *kunkum*, camphor and incense, flowers, bananas, coconuts, fresh mangos, a new sari and blouse, and red bangles, a color she enjoys. They interact with Ammavaru as with a living person, a vital presence, a human- divine being with whom they have an intimate, touching relationship. This is a family get-together as the Nayudu women lay out a feast on stitched leaves (*maḍapa*). On each leaf are placed lentils (*pappu-java*), *paramannam* rice with milk and jaggery, *pongaḍālu* sweets, *mulagaku* curry with *telaga-pindi* (night-shade and crushed sesame seed-husk, left over after crushing the seed). They say that this curry and sesame dish is a special favorite of Ammavaru. With warmth and affection, the three married Nayudu women put *pasupu* and *kunkum* on Ammavaru's *tali* and then touch their own *talis*, enhancing the longevity of their married life (*saubhagyam*). The women now burn the *sambrani* incense on a brazier.[47] Dense smoke fills the room; nothing is visible through the haze apart from two hanging lamps.

M. V. Krisnayya describes his response during those moments as one of total ecstasy. A scholar of Western philosophy, he says that he was utterly beyond its logic. All reservations left him, and distinctions between men and women, young and old, disappeared, as did all negative emotions. Within Ammavaru's home the thickness of the incense plunges the participants into the reality of her interiority, where she is (wherever that is)—into the great density of her cosmos with its continuous flow, where rupture and difference are absent and the holism of togetherness (of that which Durkheim called effervescence) fills all space/time. It is from this perhaps infinite self-interiority that ancient Ammavaru must be persuaded to emerge into the human world, to be born there as a vulnerable infant, to put herself in the hands of her human kindred, the family of Nayudus, one of whom will become in the course of her evolution the Raja, her brother.

The women return to the main Temple hall where they perform pujas to the other two goddesses whose shrines are inside—Durgalamma and Mutyalamma—and then at the margosa tree in the back courtyard where until recently a third goddess had resided. After some fifteen minutes the cloud of smoke dissipates and the women re-enter Paiditalli's room. Now

[47] This is guggulam, the resin of *shorea rebusta*. Paiditalli loves its smell.

they pray out loud to her, appeal to her at length, their message being, 'We are bringing you next Tuesday, please come (*miru tappakunda raavaala*), take care of us, protect us.' All the preparations will be made to receive her, but the choice to come is hers and hers alone. If she comes, she will enter this nexus of kindred, just as she initially left the human world when her most powerful kin tie shattered with the death of her brother, the king. The Nayudu women are asking Paiditalli to entrust her unformed, vulnerable baby-self to the Nayudu family whose head is her chief priest in the Square Temple.

The resistance to appearing within the waters of the lake that the Goddess sometimes demonstrates indicates that this may not be an easy birth, that the rupture at this time between the watery depths of female deity and the flat, dry, exoskeleton of the human world is profound; this rupture may even be disturbing to her, even though this region is part of her cosmos. In our understanding, self-birthing presumes that the Goddess is conscious of this tear; that is, conscious of human beings as within her yet as exterior to her deeper selfness, so that emerging on to the surface of this brittle, fragile portion of her cosmos may diminish her. In any event, self-birthing assumes self-consciousness. Otherwise, how would she know that parts of her must begin their evolution on the hard dryness of the human world? If she appears, she will emerge at the bottom of the very lake in which she drowned herself as the king's younger sister. There are no touchstones here, only liquid loci.

The Telugu term *karya-kramalu* recurs often during the coming events. The term refers to what we would probably call 'ritual'; literally, it means 'sequences of things that have to be done', that is, acts/actions to be done in the proper order so that they will be efficacious in accomplishing or achieving what the 'ritual' is intended to do in terms of transforming cosmos. These sequences thread through and thread together Paiditalli's coming, maturation, harvesting, and departure.

One week after the Nayudu women's visit to the Vanamgudi, the preparations for bringing Paiditalli are in full swing. The various elements and aspects of the Goddess are assembling on an early Monday evening. The locus is a large empty lot of some two hundred square yards, hemmed in by houses on all sides, accessible by a dark narrow lane that abuts on open sewers, with the low eaves of house roofs almost covering it, only a few minutes' walk from Bairagi Nayudu's home. He owns the space. This is the *Ammavari sthanam*, the Goddess Place, her special space, indeed her home in this neighborhood of Hukumpeta. The Goddess Place is now well-lit; music—a *suprabhatam*, a morning wake-up song for Paiditalli—

16. Paiditalli Pots in Goddess Place, Night Preceding Devara Pandaga (David Shulman).

blares. The space is full of people, numerous women with their children in tow, many unmarried girls, and a few men. The atmosphere is happy, vibrant, peaceful.

In the auspicious north-east corner, on a *saduru*, a low, square cement platform with a low wall around three of its sides, are eight metal pots in three rows. The cement is covered with red cloth, and *muggulu* are painted on the ground in front of the pots and along the narrow top of the low wall. The pots have been in storage at the Vanamgudi since Paiditalli left the Cadurugudi after the Sirimanu of the previous year. Also present is the *utsava murti*, the moveable *jatra* image of the goddess from the Vanamgudi Temple. The pots are a major presence during the Sirimanu ritual complex. These pots *are* Paiditalli, the containers of Paiditalli, the multiple modalities of Paiditalli, the wombs of Paiditalli, the continuous presence of Paiditalli in the human world. In the back row are two large five-metal (*pancaloha*) pots, each with a cover in the shape of a cone (*garagalu*).[48] These two are

[48] The metals are gold, silver, copper, brass, and bronze. The eight pots are also six plus one doubled, in other words, seven, that corresponds to the presence of the Goddess as seven sister goddesses, a prominent presence in this area of India.

the 'main pots', called Ammavaru Ghatalu, the pots of our Mother (Paiditalli). These two are the Goddess doubled. A goddess does not go around by herself, but always with another, and the two are understood as sisters. These manifestations are not, perhaps never, the ultimate being of the Goddess (her *visvarupa*), within her infinite depths, but rather forms closer to her surface that nonetheless are the Goddess in every respect. In the middle row there are four smaller brass pots, also with cone-shaped covers; and in the first row, two middle-size brass pots, cone-less but with plates with lit wicks covering their openings, their exteriors coated with turmeric and with many *aksintalu*, grains of uncooked rice. In front of them sits the *utsava murti* of Paiditalli.

The pots are the infrastructure of the goddess in the human world, used now to show the goddess just how dearly she is respected, desired, loved, and needed. From her distant though vibrant quiescence she is being brought, slowly, easily, intimately, back into the human world, among kindred and friends, and made to realize on some level of consciousness just how adored and valuable she is to people. The women crowd around this corner, briskly arranging and adorning the pots with deft movements. The pots and *utsava murti* have come first to Nayudu's home, where a lustration *(harati)* was performed for them, and from there into this special neighborhood space. This journey is also an intimate visit in return for that made a week before to Paiditalli at the Vanamgudi. The women present are of the neighborhood. They are dressed festively, fresh from their bath, wearing new or newly pressed saris, flowers in their hair, *bottu* on their foreheads. The married ones have come for *perantam*, a visit of auspicious, fertile women who come to bless and to receive blessing. They clutch their children as they approach the goddess. They touch the goddess, the pots, the feet of Nayudu, making their offerings to her of food, poured into the pots in the front row; bringing new saris, blouses, bangles, to show her and to have her bless them. The tone is intimate, emotional, exuberant, joyful. Their husbands are remote, keeping their distance, behaving with a sense of modesty as befits a man before a woman. The children shout, jump, play. The heavy heat blankets everyone.

Three bards, *Jamukulavallu*, arrive and seat themselves. They are in high spirits, full of smiles, though sleep-deprived and tired—this is the goddess-ritual season and they are in high demand. They sing Paiditalli's song in an archaic, very rustic Telugu, almost impossible to understand. For Nayudu, his family and neighbors, the focus now shifts to his home where a festive meal begins, to which ritualists, bards, drummers, and relatives are invited. During these hours of feeding, the women continue to throng to the Goddess Place, visiting Paiditalli. From there, in the later evening, the pots will be processed to the Cadurugudi, where they will remain until after the Sirimanu in October, and where the newborn Goddess should arrive in the early morning hours of the morrow.

By 11 P.M. the drumming increases, the police are in evidence, and the first guises, *veshalu*—of tribal Koya dancers—appear in the neighborhood lanes. In the Goddess Place the pots have been lifted onto peoples' heads. The entourage slowly emerges from the enclosure, through the narrow defile, into the street. Women line the doorways, expectant, excited; many reach out to touch the pots that together with the festival image are at the back of the procession. Men dominate the procession, and only men dance. At the front are the *veshalu*—tigers, Koyas, transvestites, stick wrestlers (*samugaridivallu*), and the guise sometimes said to be Paiditalli's own true form, the *Kalika-vesham*, usually meant to be wrathful, frightening, unpredictable, suddenly appearing and disappearing. It has long black hair, a peacock-feather headdress, a trident resting on one shoulder with green limes speared onto its tines, a skull on the other shoulder, a red-cloth breastplate, a green skirt, bells tied to the guiser's ankles, and green limes in his hand. There are also free-lance dancers (*janaka janakalu*), dramatic in their movements, bounding and leaping through the streets to the beat of the drums, intended to enhance the enjoyment of the Goddess.

In the middle of the procession is a ragamuffin band—a clarinetist, a classically trained Nagaswaram-player, and a hurdy-gurdy harmonium mounted on a cart—filling the night with music. Bairagi Nayudu, who has been holding the *utsava murti* of the Goddess, passes it to his younger son, Venkat. There are perhaps a hundred people altogether in the procession, but the atmosphere is electric, celebratory. The happiness in the more familial Goddess Place now spills and flows through the streets, as the procession moves very slowly, meandering carefully through the narrow lanes, stopping almost house by house so that the householders can greet the Goddess. There are more police, and more of higher ranks, monitoring the movement. A few isolated raindrops fall, the air is suddenly cooler, and a sweet wind fills the streets, the sky full of lightening flashes. Everything is being done to encourage Paiditalli to come into the human world.

About 2 AM the procession reaches the Cadurugudi, and the pots and image are taken inside and installed in the *garbha gṛha*, the inner sanctum of the deity, with its permanent image, the *mulavirat*. The door to the *garbha gṛha* is locked, as is the temple door. The Goddess pots await within the coming of the newborn. Outside the temple the main thoroughfare, Three-Lantern junction, looks like a war zone in the darkness empty of people; the day-time crust of human beings in continuous, uneven motion has been peeled away, revealing the entire length of the main road as a chaotic jumble of potholes, rubble, loose piles of bricks, collapsing facades, mounds of dust, broken steps, all swathed in the grime of sweat, doggedly and grimly mysterious.

Around 2:30 AM a small procession sets out from the Cadurugudi for Big Lake. But instead of the pots there is a winnowing basket (*jangidi*) carried on the head of Paidiraju, a Talaiyari (oil-presser) and a great servant of the Goddess. A red canopy bordered in orange, its middle pushed upward with a long stick, is held over him by six men, a makeshift roofed dwelling in motion. In the procession of perhaps a hundred are the drummers, the bards, many more Harijan-Malas and others of low caste, with the noticeable presence of women. This time the procession goes directly to the lake, its ebony waters dotted near the shore with algae and lotus pods, arriving shortly after 3 AM. Along the length of the lake are many small clusters of people waiting, again many women and Harijans (Dalits) who frequently attend the Cadurugudi and who have a special relationship to the Goddess. The location chosen for the Devara pandaga ritual is about halfway between the Cadurugudi and the Kota on one side of the lake with the Vanamgudi on its other side.

As the drumming intensifies the *jangidi* is brought down to the lake along a narrow footpath and placed on the slippery muck a few inches from the water's edge. The women and Harijans take control of the rite. Paidiraju's wife cleans the ground of debris, and dry palmyra leaves are collected, to make a fire for *harati* after the goddess arrives. Paidiraju's brother digs a hole in the mud at the water's edge. Some dozens of people gather around the *jangidi*. The basket's concave inner surface has been rubbed intensively with *pasupu*. In the center of the basket is a largish circular bed of *kunkum* surrounded by white flowers. On the *kunkum* bed is a circular metal vessel filled with camphor, in which is a long, lit wick and a raw mango. The whole *jangidi* is formed as female. The winnowing basket is rubbed with *pasupu*, as is a woman's face in process of intensification. The basket is marked with the *bottu* of *kunkum*, as is the female forehead. In Andhra the winnowing basket is also strongly associated with the womb and female fertility (Handelman, Chapter Three), and the mango with the vagina and the birth of goddesses. Thus: face within womb, vagina within the face, a lit lamp on the forehead, a mark of respect and worship. The female turns into and through herself, interior becoming exterior, exterior becoming interior. The goddess gives birth to herself, first in the lake and then onshore, at the water's edge, in the *jangidi*.

17. Awaiting the Birth: The Jangidi at the Big Lake (David Shulman).

We note the dynamic condition of the Goddess. If she comes, emerging from within herself to the lip of the human world, she will then implode once more into self-interiorization. If she comes, the newborn amorphous infant (of mud) will be placed simultaneously within the female form (the *jangidi*) yet also on its intensified (and therefore deeper) surface, from which she will continue to emerge and mature within the human world. Perhaps we can say that, coming from deep within herself, she will be placed deep within herself on the shore of the waters of wilderness; thus she will emerge from within herself in a two-fold dynamic, phase by phase. First, she comes out of the fractal depths of wilderness to reach its shore; second, she enters into the human forming of the female that is the *jangidi,* still on the shores of wilderness. Later she will emerge further in the Square Temple of the *uru,* in the crux of civilization. Earlier we referred to the entire dynamic as one of parthenogenesis ('virgin creation'), self-birth, self-generation, self-creation. In the relationship between lake and *jangidi* and, too, within the basket itself, Paiditalli is at once inside and outside herself, and thus she gives birth to herself without ever leaving herself. In this sense she shifts modalities of selfness, permitting human beings to shape her in relation to their need to create life and depth within the flatness of civilization.

What occurs within Big Lake? On an earlier occasion, Bairagi Nayudu told us that he stands in the water, entreating the Goddess to come. This ritual process can be difficult and may take hours of begging, coaxing, cajoling, and scolding the Goddess, telling her she is disinterested in the welfare of her people, and the like. If all else fails, they curse her, arousing her awareness of them but also her anger, with powerful effects on those who await her. Eventually a light, a spark, or the end point of a lightning flash, coming from the direction of the Wilderness Temple, will appear in the dark sky, fall into the water and come to rest in the mud, its colors those of golden turmeric and vermilion. This is Paiditalli, the golden goddess, coming with great force, knocking consciousness out of Bairagi Nayudu and from others standing in the lake as they grasp the mud on the lake bottom. In 2006 he described what happens somewhat differently: 'We go into it [the lake]. We take mud from it. A spark will come into my hands. I will be like this [his hands are cupped; he already has scooped up the mud and holds it in his hands, awaiting the strike of the spark]. Then I close my hands [over the spark]. In my hands there will [now] be [the colors of] turmeric and vermilion. If not, that means the goddess has not come.'

In May 2006, Bairagu Nayudu is too unwell to enter the water and rests in the ethnographers' car. His place is taken by his son-in-law, Srinivasa Rao, a priest in the Wilderness Temple. Srinivasa Rao and three other men—Venkat (Bairagi Nayudu's younger son), the Talaiyari, Paidi Raju, and the Mala, Appa Ravu—enter the water, accompanied by drummers and the *Jamukulavallu* bards. They wade some 150 meters in, the water reaching their waists. They stand randomly but quite close to one another. Some 50 meters behind them another group stands in ankle-deep water—mostly young men and boys, helpers and onlookers, ready to carry the four out of the water when they lose consciousness at Paiditalli's arrival. Both groups face east, toward the Wilderness Temple diagonal to them on their right, across the lake. On the shore the Paidi Raju family women and a few others huddle around the *jangidi*. Behind the winnowing basket are two of the goddess's great acolytes, Satyanarayana and Bhaskar Ravu, seated on the soggy ground. In a half-circle around them stand another thirty or so onlookers. Farther back, closer to the road, are more groups of spectators, strung along the lake front. In the car on the ribbon of road, Bairagi Nayudu waits. The sky is alive with lightning, the air deliciously cool, the burning heat of the dry season forgotten.

The night is quiet, as all wait in the gentle silence (so unusual in South India). The backs of the four are dimly visible in the black waters, against the dark sky lit by fiery flashes. The mood on shore is easy, relaxed. There is a subdued sense of anxiety, though people are mostly happy, serious, expectant, waiting with patience as the minutes pass. A deep serenity pervades the scene, a softness quite unlike the intense nervousness and frenetic activity of the Sirimanu Jatra, which marks the climax of Paiditalli's growth and maturation in the human world. Once in a while a ripple of movement is visible among the men deep in the lake. More minutes pass. The water below seems to join with the sky above. Suddenly the boys in the water are running, crying, '*Occestunnaru*' [They are coming!].' They have jumped the gun. More waiting. David Shulman, watching from the shore, has the feeling that he is present at the birth of time, time that will open to swallow everyone. But as the Goddess emerges from the watery depths, bringing wilderness into civilization, she also speeds up time, creating a time of growth and ripening of the fruits-to-come. This faster time is continuous, with fewer breaks, fewer of the stops and starts that are the hallmark of civilization and its organization. So, too, the spaces the Goddess will move through will become more continuous with one another, brought together, *intra-connected,* with a lessening of boundary markers and internal divisions. Time and space will be in-fused with the liquid qualities of Paiditalli's presence. Her coming will join together four of the five basic elements of which the Hindu cosmos is constituted—fire (the lightning), air (through which she will pass), water (into which she will plunge), earth (into which she will enter on the lake bottom).

Again there is movement in the water. The boys standing in the rear rush forward, the small group of men disappear into the water, and a series of soft cries pierce the darkness: 'Jai Paidimamba.' There is splashing, a flurry of movement. Srinivasa Rao is dredged out of the water, his cupped hands heavy with amorphous greyish-black clay mud. He is unconscious. Then Paidi Raju, Venkat, and Appa Ravu are dragged onto the shore, all drenched, muddy, beyond communication, their eyes shut, their hands clasped tightly around clay. The Goddess has arrived with force and been brought to shore, one of the men says, by Srinivasa Rao. People force the men to sit up, and gradually they return to themselves, opening their eyes. As if delivering an infant, the women, working quickly and efficiently, take the mud from Srinivasa Rao's hands with infinite tenderness into the *jangidi,* cleaning and washing, making order, mixing the goddess (who has arrived with her own turmeric and vermilion) into the turmeric and vermilion within the winnowing basket, kneading, molding and feeding her within this exterior womb on the surface of the human world, an exterior womb that is no less a cradle.

While a puja (of the variety performed for village goddesses) is being performed for Paiditalli, the women and Satyanarayana cover the winnowing basket—the amorphous infant within—with their hands. The women, particularly Paidiraju's wife, perform a secret ritual over the infant. *Kunkum* is applied to a pile of white rice, turning it red, so that it looks like the blood-rice of a *bali* sacrifice. One corner of the basket is opened, and an egg, an onion, and other food are placed inside to feed the infant.[49] Various substances are passed quickly from hand to hand. Seven raw eggs, other foods, turmeric and vermilion, are buried in the hole dug earlier by the water's edge. This, it is said, is to ask forgiveness of Paiditalli in case any mistakes were made. The winnowing basket is placed atop this covered hole.[50] A fistful of mud is taken from the basket, a lit oil lamp placed on it, the mud then placed atop the basket. Again, as in the growing of Taravani, this ritual is spoken of as 'the making of *layers*', that is, the self-intensification and focusing of presence, the opening of depth and space within the infant on the flatness of the human world. In the ritual the elements of the goddess are placed under the earth (deep in the hole), in the (deep) womb of the *jangidi,* and (into the depths of height) atop this winnowing basket. So, too, the infant is nourished on each of these layers. Paiditalli, coming as water and earth, will evolve and expand within human flatness, deepening it, opening its earth to the rain waters, making it fertile.[51] During the puja, people stand around and talk of the coming of the goddess. This year, they say, she has come quickly and easily, 'without any trouble'. She is very innocent, they say, 'a cool mother'—innocent

[49] The food is simple peasant fare of the kind used to feed village deities.

[50] Nonetheless, the number seven, here intimately related to Paiditalli's re-birth, reminds strongly of the seven sister goddesses, of whom Paiditalli is at times one.

[51] The food and substances placed within the jangidi and those placed in the hole, then covered by the jangidi, seem to be quite similar—close to being mirror-images, one entering Paiditalli's limitless interior, the other entering the human world. We could extrapolate here that the 'making of layers' is literally the making of a simultaneously present three-tier cosmos as Paiditalli emerges from the waters. The lowest level is beneath the surface, propitiating and feeding her within the mud before she emerges onto the surface. The mid-level is that of the infant Paiditalli who has emerged onto the flat surface and who is being fed by the water's edge. The upper level (the fist-sized bundle of mud topped with a lit lamp atop the jangidi basket) is higher, perhaps a dimension of futurity that will come into play with Tree-Paiditalli. This rite, then, incorporates the simultaneous existence of these domains and moves between depth and height, as does Paiditalli during her ritual presence in Vizianagaram.

in the sense of being unheated by the experiences and influences of the human world. Let us call her in this form "Mud-Paiditalli."

The first bird-cries of dawn, a rush of wings, the cock's crow, people stretch and gather themselves; the procession is prepared again. Bards and drummers to the front, then Paidiraju carrying the *jangidi* on his head under the red canopy (now a baby dwelling), they start on their way to the Cadurugudi, the Square Temple in the *uru*, the Old City. The route is specified: certain lanes are important, and there people make traditional offerings. Saris are spread on the road, all the way to the temple, so that the goddess does not touch the ground. As the procession passes through neighborhoods, mothers come into the streets and lanes with their small children, prostrating in front of Paidiraju, making him step over their little ones. When the *jangidi* reaches the Square Temple, Mud-Paiditalli is placed on a mat next to her little shrine tucked into the three-species tree, outside and in back of the Square Temple. Some say this is the original square mat on which she sat long ago next to the tree, before her tiny shrine was built.

The mud is brought into the *garbha gṛha* of the Cadurugudi, the Square Temple. A small amount is placed at the feet of the Mulavirat, the permanent image of Paiditalli. The rest of the mud is spread on the *capa*-mat in two clumps, to act as 'cushions' (*kuduru*) for the goddess who comes as the two five-metal pots (*talli-ghatalu*). On each side of these two is placed one of the two largest metal pots. And on each side of these two are placed two of the small pots. Some mud is placed under all of these pots. The pots sit like this in the *garbha gṛha* for fifteen days. In addition, a lamp is lit which must burn for an entire twelve months until Mud-Paiditalli returns.

On the fifteenth day (of course a Tuesday) all the mud, now exhausted and devoid of the pasuppu and kunkum energy of Paititalli, is carefully scraped up and put back into Big Lake. The area in which the pots stood in the *garbha gṛha* is then ritually cleansed. On that same Tuesday, two new pots (*kuti-ghaṭālu*) made of a particularly soft clay (*liguru-maṭṭi*), such as one finds on the bottom of Big Lake in particular spots or in a riverbed, are brought into the *garbha gṛha*. They too come in twos and are meant to collect food for the goddess from neighborhood households.

For the next months, through August, the infant Paiditalli, amorphous in form, matures in the human world, transferring her essence, her energy, to the pots, to this more definite, intense, focused form of goddess, as the pots regenerate, soaking up her youthful presence. Yet in an important sense, the spheroid pots too are her own womb within which she grows, as the *jangidi* was her womb, as was Big Lake itself. She continues to grow, to develop within herself, evolving her layering and her complexity, even as she moves from womb to womb. She shifts into wombs that are also products of civilization more compatible with the transformative tasks she has embraced here—such as the task of infusing the dry, shallow land with her inherent fluidity. She is the golden goddess and, as we have said, gold is the solid closest to the fluid. Even as her form acquires a measure of solid presence, she continues flowing with herself. In this regard she has flowed to the Square Temple, there to flow into herself as pots that will carry her pregnant fluidity into the city, deepening and filling its arid cysts of cosmos, sowing them, preparing them for harvest.

Paiditalli comes out of the waters, emerging from herself to become watery earth on land. Even though she came easily in 2006, more usually persuading her to come forth requires great effort. She may well not want to come, not want to leave the depths of cosmos where she is most at one with herself, deep within herself. There, self-consciousness as this is understood in the human world may not exist. Praising and cursing her, singing her qualities, telling her story, are all attempts to awaken her to herself, to self-recognition, by awakening her to the existence of human beings. Eventually she comes for the sake of her people, yet she does so with great force, itself often understood as *anger*, striking unconscious the priest and his helpers whose task it is to remove her from water to flat land. Moreover, her coming is the force of creation, always the re-forming of form. She appears in the depths of herself as primordial mud infused with her essence of golden yellow pasupu and vermilion kunkum, quintessentially female in its first moments of self-awareness, awakened as she enters the watery depths.

There seems to be a change in her being as she arrives. Coming with force, perhaps fury, she seems fully adult. Yet becoming mud on the lake bottom, she becomes quiescent, more like a slumbering infant. As soon as she is onshore she is worshiped. Yet she is not mature—perhaps "evolved" is a better term, relating to awareness—and will not be for a long time. She is described as 'innocent', as prepubescent, her malleable form amorphous, placed in the womb-like, cradle-like winnowing basket. Her coming echoes the myth of Paidimamba who returned from these same waters, this same womb, as the protectress of kingdom and kingship and therefore of the kingdom's own growth and maturation that is linked inextricably to agricultural growth and maturation, especially that of rice.[52]

---

[52] In some sense too, the priest and his helpers 'die' in the waters, losing consciousness, returning to life, to consciousness, on the banks of Big Lake, on the edge of wilderness.

The configuration of male seed germinating in female earth is widely present in South Asian agriculture (Balzani 2003: 159). Stutley and Stutley (1977: 278; see also Balzani 2003: 162–165) citing the *Taittirīya Saṃhitā* v. 2, 5, comment that in ancient India, 'at the beginning of each sowing season, twelve furrows representing the months of the year, were ritually plowed by the king or chieftain in the first field…As the repository of the seed, furrows were regarded as the earth's female generative organs.' Indeed, 'the word bija may refer to both seed and semen' (Balzani 2003: 157). In coastal Orissa and likely in northern Kalinga, ploughing and sowing the earth is homologous with sexual intercourse (see Apffel-Marglin 1981: 160).

6. *The Growing of Rice*

Rice was the main crop of the kingdom of Vizianagaram, a staple that must have timely and sufficient water. The major source of water is the great southwest monsoon that should end the dry season, usually arriving around mid-June (Mṛgasira month). This is when rice is planted. The periodicity of cultivating rice is intimately related to the cycle of Paiditalli's self-birth and evolution. Farmers in villages around Vizianagaram speak of rice in the language of a woman growing to sexual maturity, becoming pregnant, maturing to parturition, and then 'drying out' after childbirth.

The rice-plot (*āku-maḍi*) must be ready before the first rains. The rice-plot is a small corner of a larger field. As Moreno (1992: 151) comments, farmers draw 'homologies between human bodies and the earth as a divine body of the goddess Bhudevi; it is clear that agriculture in India has the characteristics of bodily sacraments (*samskara*) and that the farmer occupies the position of custodian of the earth's potential for "reproductive prosperity" (*sri*).' The land is plowed a number of times and the furrows manured.[53] Around early July, after at least a few days of rain, so that one can step into the water in the field up to one's ankle, the rice-plot

[53] Commonly in the Hindu world, plowing and sowing are understood as sexual acts through which Bhu Devi is impregnated—while, for example, the transplantation of rice seedlings may be thought of as 'merely altering the seedlings' location within the already pregnant goddess's womb' (Daryn 2006: 195). Daryn (2006: 194–219) has an especially insightful discussion of how among Hindu farmers in central Nepal this pregnancy develops to fruition.

In the villages of Orissa, indeed in North Kalinga, during the ritual time of *Raja Samkranti*, the earth, that is, Bhudevi, menstruates; and during this period (sic.) farmers refrain from any action that would cut the earth (Apffel-Marglin 1981: 161). But Bhudevi menstruates in

is planted with seed. After the paddy seeds are spread, the land is flattened with a board and covered with thorny bushes to prevent the cattle from wandering over the surface. After six days the first tiny white sprouts (*molaka*), three or four inches long, appear. The sprouts continue to grow for another fifteen days or so, turning bright green in color. Then, usually at the end of July, sometimes in early August, the sprouts are removed, made into bundles, and transplanted (*dammu*) into the larger field.[54] Commonly, the farmer's wife will do the first transplanting in the auspicious north-east corner of the field—also the last transplanting (unless she is menstruating, in which case the farmer does it).

Planting and transplanting are folded into local rituals of fertility and protection. Before the transplanting, the farmer takes the farm animals to an anthill in the fields and performs a puja there to a local goddess for the well-being and fertility of the animals. Sometimes blood-offerings (of chicken) are made. The event is called the Cattle Festival (*Pasuvula pandaga*). With the first transplanting, the farmer's wife performs a puja to Bhudevi, who is ultimately responsible for the growth of the rice. This is followed by the women's worshiping the village goddess (the *grama devata*), asking that crops grow in abundance. When all the farmers of the village have completed their transplantation ('not even one left behind'), the village as a whole celebrates the Village Deity Festival (*grama-devata-pandaga*). Even as the first shoots are grown and then transplanted, Mud-Paiditalli is germinating in the *garbha gṛha* of the Square Temple, transferring (perhaps one can say 'transplanting') herself from mud into pots, her presence evolving in the human world, from womb to womb. This language of germination, *vittanam nāṭṭu*, is common to rice and to sexual activity—a language infused with the feeling of sexual dynamics.[55] For example, the sensuality of thrust ('Did you feel the thrust?' [*bāgā nāṭṭindā*]) and the proprieties of timing (for the wedding, for first sexual activity) apply equally to rice cultivation. So, too, after the seeds are spread in the north-east corner, the farmer watches carefully, day by day, anxious to know whether they are good enough—just as after the wedding everyone is anxious to see if the seed sprouts, if the new wife becomes pregnant. The farmer must see to it that the depth of water in the furrows is

---

the 'hinge' between the dry and the rainy seasons, showing her fertileness as the agricultural rhythm thrums from barrenness to sowing and growing (Apffel-Marglin 2008: 21).

[54] The Telugu word for transplantation, *siddham ceyaḍam*, means making something ready, fine-tuning it.

[55] See also Gold 2003: 259.

appropriate, not too much, not too little, around two to three centimeters. Water flows from one plot to another, but each rice-plot has water gates to save or drain water, and levels have to be continually adjusted.

The first tangible, visible stage after transplantation is called 'becoming a stomach' (*poṭṭa rāvaḍam*), as pannicle buds (called the 'little stomach') start forming inside each of the shoots or sheaths of the plant. This is akin to the beginnings of female pregnancy. Within seven to ten days, as the pannicle bud swells into maturity, becoming the 'stomach' (*poṭṭa*) emerging from its sheath, the flower (*pūvu*) is visible. By late August, as the stomach is growing, the new clay pots, full of Paiditalli, begin making their daily rounds from the Square Temple into the lanes and streets of Old Vizianagaram.

The dynamic of flowering is internal, secretive, unseen. The flower is like a 'soft downiness' (*nuguru*), 'a hairy covering of leaves on stalks of some plants' (Gwynn's Telugu-English Dictionary). In Telugu, this flower (*pūvu, pūvāram*) means a compact tenderness that has to transform further. It has *pūta* in it, a kind of painting on, rubbing on, applying. In this there may be intimations also of intensifying, perhaps of layering. During the time of flowering, the plant's male and female reproductive organs are pollinated together by the wind. After a further seven to ten days, the flower turns into seed (*ginja*), and when a bunch of seeds come tightly together, the farmers call this the backbone (*vennu*). The fertilized flower develops a quasi-protective hull that fills with liquid starch and protein, and as the flower falls away this milky fluid becomes visible. This is the most significant stage for the farmers, proof that the soft, green seed is a viable offspring. They call this stage the milk-pregnancy (*pāl-posukovaḍam*). 'The flower becomes pregnant with milk'; literally, 'the flower pours itself milk [within itself]' (*pūvu pālu posu kondi*). A standard phrase used for women's pregnancy is "She has poured water for herself" (*āviḍa nīḷḷu posu kondi*).

The parallels between these pregnancies are vivid. The plant is successfully pregnant within itself as its seeds develop; it then turns hard and yellow (*pasupu*), ripe with maturity (*paṇḍe tappadu*), the shoots heavy with rice bending back and down (*ongaḍam/vangaḍam*), ready for harvesting. The significance of the milk-pregnancy is underscored by the Stomach-offering sacrifice, the offering for the success of the 'pregnancy' of the rice plant, that may be held at this time: a *potu* male-sheep is slaughtered, its blood smeared on cooked rice offered to the goddess in the hope that nature will be cooperative. The crop is not harvested until it has ripened itself, cooked itself, to the *pākam* moment, to the right point. Farmers may check their fields three times a day for this right moment, when the liquid in the hull hardens to form a starchy inner grain. As soon as this moment is reached, they sacrifice chickens to Bhu Devi and harvest the crop. The first cut—'holding the fistful' (*piḍi anṭaḍam*)—is offered to Bhu Devi and kept in the north-east corner of the field. Usually, harvesting is in November, in the month of Kartika, some four months from the spreading of paddy seeds in that auspicious north-east corner. After harvesting, the paddy has to dry for one week (*eṇḍāli*) before it is brought to the threshing floor and, following that, winnowed. In parallel terms, after delivering her infant, the mother has to be dried out (*ārāli*). She is called raw, uncooked, undried (*bālintālu*). She is dried out with special foods (for example, preparations of betel leaves [*kāyam*]) that generate body heat.[56]

The annual rice cycle in the region of Vizianagaram is related intimately to the opening of space and depth—in seeding, the extrusion of shoots, ploughing furrows, filling them with water, transplanting, the protrusion of the stomach in the extruding plant, the forming of the milk-pregnancy, the birth of the mature *pasupu*-colored rice. The dynamics are those of interiority, of depth, exteriorizing itself through emergence onto the softened, now-receptive, indeed welcoming surface of the human world. As we have noted, it is within the space of opening and the fullness of depth that desire is profound and change is made. The dynamics of exteriorization are primarily female, a matter of generating new life from within life. The generation of rice and the generation of life through female pregnancy are, as we have said, intimately related.

Yet more powerfully than this, once again the parallels seem to suggest a very high degree of fit in this cosmos. To say this is to emphasize the depth and breadth of homology among different domains—the emergence of the goddess, the emergence of the human child, the emergence of rice—so that these domains can be said to emerge from one another, just as we argued earlier for the relation between water and mountain, one emerging from and turning into another. So, too, these domains fold into one another to a large degree. The goddess emerging is an infant, as is the human child, as is the newly born rice before it is processed. All undergo processes of maturation within the human world that make

56 Among Tamil rice-farmers, the traditional Tamil New Year of Tai Ponkal is marked by the cooking of raw ponkal rice that in these moments of transformation and passage from one year to the next, the opening of cosmic space, condenses the entire dynamics of the rice-growing cycle, from planting to harvest, within the cooking pot-womb of Bhu Devi (Good 1983: 236).

them fit and able members of *this* portion of cosmos and that shape their capacities for particular interactions and accomplishments.

7. *Pot-Paiditalli Furrows and Sows*

During the period that the stomach, the flower, and the milk-pregnancy appear in the rice plant, the two *kundalu* clay pots of the evolving Paiditalli begin to move from their location in the *garbha gṛha* of the Square Temple. During the evenings of August and September the pots go in procession into Old Vizianagaram. Let us call this shaping of the Goddess "Pot-Paiditalli." The Old City is built according to the form of a square *mandala* with thirty-six streets to a side.[57] Every Tuesday evening during this period Pot-Paiditalli goes to visit the royal Śakti, sometimes identified with Kanaka (Golden) Durga,[58] who seems never to leave the depths of the Rani's quarters in the Moti Mahal, or the Round Mahal, of the Kota palace-fort and perhaps the Durbar Hall in the adjoining building, where court rituals were held.[59] On Thursday evenings during these weeks the pots go to the Goddess Place in Hukumpeta, there to be worshiped by the women of Bairagi Nayudu's family and the women of the neighborhood. On other evenings the pots follow a set pattern that during these months will eventually take Pot-Paiditalli into many or most of the streets, lanes, and byways of the capital, into its nooks and crannies as she goes from home to home, to be met by family members, especially women, making their offerings and asking for her blessing. This worship night after night, street after street, is akin to *furrowing* the flatlands of the city, opening space for the depth of presence of the goddess within home after home. So, too, one can think of these movements of the goddess as *sowing* the coming of kingship in every corner of the furrowed *uru*. The climactic harvesting will occur during the Sirimanu, as the king is brought anew to his palace, thereby renewing this integrative core of the kingdom.

There are, then, three vectors through which, during this period, Pot-Paiditalli opens, sows, and energizes the city: that of kingship in her journeying to the Śakti of the Kota; that of kindred, in her journeying to the

---

[57] This is the mandala in abstraction. On the ground, the Old City has far less than the 1296 streets the mandala would generate.

[58] In South India, and in Kalinga, Durga is the great protectress of kingship. See Schnepel 2002: 255–292; Dirks 1987: 39–40, 285 f.; Waghorne 1994: 215.

[59] The Round Mahal, indeed round in circumference, may have taken its name from the royal marriage bed, which was perhaps round in shape.

Goddess Place in Hukumpeta; and that of kingdom, in her journeying through the byways of the Old City. Though kingship and kingdom were abolished in 1949, their presence is integral to Paiditalli and the Sirimanu. In practice, royalty is an almost-presence, its niches embedded in the rituals and in the teleologies and efficacy of their practice, yet rarely if ever actualized. The primary royals are the two sons of the last Raja. The eldest, Anand, the Raja-who-would-have-been, is treated as the Raja. He is always expected to appear at Paiditalli rites yet rarely does so in person, though sometimes he sends a representative; and he makes the gifts expected of him in his royal capacity. Though he has a great house in the city, its appearance is that of shuttered abandonment, its yards hardly cared for. Introverted, private, he lives much of the time in a modest home in the mountains, within sight of the great temple of Simhachalam.[60] Nonetheless, it is crucial to note that in all the Paiditalli rites the king himself is not the focus. Paiditalli's re-birth and relationship to the kingdom is the focus. The king's presence should touch these rites at certain points, yet the person of the king is not affected by them. We return to this in the conclusion.

The Tuesday visits of the *kundalu* pots to the Kota Śakti during this period of deepening the city with the blessings of Paiditalli are crucial to this ritual complex: they mark the venue for harvesting kingship within the furrowed kingdom. The Indian State abolished Vizianagaram kingship, but the Kota Sakti has never left the palace-fort. Were there still a kingdom, its deepest interiority would be in the palace-fort. In the Kota the Śakti's home is within the Moti Mahal, the quarters of the Raja's consort, the Rani, and in the Durbar Hall. Traditionally the Gajapati kings (including the last Raja) were crowned in the Moti Mahal. Not only has the Kota Śakti not deserted the palace-fort, to our knowledge she never goes out of this interiority of the kingdom-that-was; to some degree she still has a part in the growing of kingdom and kingship that the ritual performs. The stable presence of the Kota Śakti attests to the powerful male presence of the palace-fort. As we have seen, Vizianagaram tradition claims that

---

[60] By contrast, in the neighboring former little kingdom of Bobbili, where the Gajapati king of Vizianagaram met his death in the eighteenth century, the rajah-who-would-have-been (now deceased) maintains a royal, or at least noble, lifestyle, as the protector, guardian, politician, and sometime provider for various populations in the area. During our brief visit to his palace, he sponsored the annual Ayudha-puja ritual for those who gave him allegiance, and he received a delegation of local tribal people who presented him with a beautiful, white, garlanded kid as a sacrificial offering (he accepted the kid, adding to us that he certainly would not sacrifice it).

the Kota itself was built on a spot recommended by the Sufi Pir Denkha Shah Wali Baba[61]—because it was a location replete with male strength.[62] According to Bairagi Nayudu, even now there is a small stone image of Kota Sakti in the Fort, perhaps in the Round Mahal, where the last Raja lived until 1958. The image is said to look like a post (*koyyabomma*). Until the end of the reign of the last Raja, the Queen Mother, Amma Sarkar, saw to it that all the rites of the Kota Sakti were carried out. The Raja-who-would-have-been, Anand, said that at night the Kota Śakti walks the walls of the palace-fort, always on guard. The identities of South Indian goddesses are at times highly specific and at other times highly amorphous. Sometimes they are quite singular; sometimes they proliferate and multiply. Sometimes hardly any distinction is drawn between the Kota Śakti, Kanaka Durga, and Paiditalli, and they are referred to as overlapping; at other times, this is not the case.

Who is the Kota Śakti? Like Paiditalli, like the generic goddess, the generic female, the Kota Śakti is depth as well as the potential power of fertility, creativity, generation and growth that comes from depth. At the middle point of the kingdom, the palace-fort, and within this point at the still deeper middle of the Moti Mahal or the Durbar Hall, the Kota Śakti is the constant presence of depth in space through time. Her very name, *śakti*, tells of her intensity and concentration of power and, if one can put it this way, her responsibility for activating the king as ruler of the kingdom. Put differently, she is female sexual fluid (*rajas*) embodying potency which is the essence of śakti (Apffel-Marglin and Hudson 2008: 83). The Kota Śakti is the female principle that activates maleness, whether male divinity or its derivative, male kingship. Apffel-Marglin (1981: 178), writing of kingship in coastal Orissa, says that, 'The power of the king is essentially a female power, a power of fertility and a power of arms. That power is called śakti…' Without the *śakti*, kingship is without creativity, without strength, without force. In this regard, *śakti* is an elemental presence where (and when) kingship relates first and foremost to itself. At Vizianagaram this presence of śakti is within the palace-fort, probably in the royal throne (*gaddi*) in the Durbar Hall, or in the cushion which supported the

61 The Pir was said to walk on air, one foot above the ground.

62 According to another story, while the foundations for the fort were being dug, the image of Paiditalli was uncovered and became the sthala murti of the Caturugudi temple. This image is small, of black basalt, her head angled to the right, with an angular face set in an expression of wonder and surprise. Thus one of her images, that in the Vanamgudi temple, came out of the water, and one, that in the Caturugudi, out of the earth.

king on the throne (see Mayer 1985; Balzani 2003: 3). In Vizianagaram this śakti would be the Kota Śakti, which may be why she never leaves the Kota. The power and protection of the Kota Śakti are ever present for royalty; and her presence does not wane. She enables kingship, yet she does not shape the annual growth and florescence of kingdom and kingship. This is Paiditalli's task, accomplished through her own annual regeneration and growth.

On the other hand, the presence and power of Kanaka Durga has generic aspects of the goddess (at times the same statement will apply to the Kota Śakti and to Paiditalli). In Vizianagaram the time of the Sirimanu is synchronized with and activated by Durga's great victory over Mahishasura, the Buffalo Demon. In this region there is a great profusion of goddess shrines, yet very few shrines of male deities.[63] Paiditalli journeys to the Kota Śakti on Tuesdays, the day of the week when the presence of Telugu goddesses is closest to the world of human beings—a creative and potentially chaotic time when cosmos may up-end itself. So, too, the Sirimanu will be held on the first Tuesday after Vijayadasami, the day celebrating Durga's triumph which takes place at the end of Navaratri, the nine nights of Durga. As Paiditalli journeys to the Kota Śakti, she strengthens their relationship, their complementary protection of kingship, thereby opening and deepening the vector of kingship in Vizianagaram by coalescing the female powers of transformation and protection. This coalescence reaches its climax in the Sirimanu, as Paiditalli carries the resurgent king, her brother, from her Square Temple to the Kota, giving his activation and care over to the protection of the Kota Śakti (perhaps especially during the lengthy period when Paiditalli will be absent, immersed within herself in the Wilderness Temple).

The interiority of the Goddess has changed since she left Big Lake in May. Mud-Paiditalli emerged from her own womb as an unaware infant in order to become her own womb again, Pot-Paiditalli, now pregnant with fertile energy. She is now, we believe, fully self-aware in the human world. Rather than her worshipers needing to attract her attention, to make her aware of them, she now goes to them daily, seeking them out in the nooks and crannies of their living spaces in order to bless and care for them. Perhaps this is her own transplantation, from her Square Temple,

---

[63] M. V. Krishnayya reports on a survey of temples in the coastal city of Visakhapatnam, some sixty kilometers southeast of Vizianagaram, in which there is a multitude of goddess shrines, yet hardly one major shrine dedicated to a male deity.

her interior, into the kingdom, household by household, each one opened up by the furrows of worship. She has shifted within herself through a series of cosmic interiorities, moving from her cosmos of utter continuity lacking all distinction toward her full presence in that portion of her cosmos that is marked by human discontinuities—deep and shallow, flowing and encysted, female and male. Thus she has moved from within her own middleness, where the self-awareness that depends on otherness is superfluous, into the womb-like depth of Big Lake; then into the womb-*jangidi* on the hard, flat surface of the human world; then into the womb-like pots at her own Chaturugudi. Now she is actively moving into the thin hardness of a kingdom in need of healing.

As Pot-Paiditalli's sowing of the city moves towards completion, she reappears elsewhere in a distinctive but no less dramatic mode, announcing her readiness to become the Sirimanu. In the second half of September, Paiditalli comes in a dream to Bairagi Nayudu. She tells him that she is growing in a tamarind tree (*cinta ceṭṭu*) outside the city, and she tells him in which direction to look for it. She tells him to come for her. For this purpose the tree will be higher and heavier than the usual tamarind.[64] Once he locates the tree, Bairagi Nayudu goes to the family, to persuade the farmer who owns the land to part with his tree. Commonly, the family resists giving up the tree, in which case the goddess turns on them in fury, bringing them disease until they acquiesce. In 2002 the goddess tree was located in a grove of tamarinds (*cinta toppu*) some two hundred meters from a road that meanders among paddy fields, about fifteen kilometers west of the city, in the blue shadow of the Eastern Ghats. The farmer whose land this was did not agree to part with his tree, but Paiditalli appeared to his mother in a dream as a snake. Frightened, the mother convinced her son to give in.[65] The tree was marked with a *boṭṭu* (*ammavari cett' ani bottu pettestam*) and awaited her devotees.

8. *The* Tevadam *Rite: Bringing the Sprouting Goddess Out of the Earth*

Since her (re)birth, the rites of Paiditalli have extended their social circumference as she emerges toward the climactic Sirimanu. Before her (re)birth, the women of the Nayudu family visited her in the Vanamgudi

---

[64] Bairagi Nayudu says the tamarind is a rare tree in this area, because of its height. The tree can grow up to 20 meters and gives fruit. Its wood is hard and dense, its heartwood dark red, its sapwood yellowish in color.

[65] He is the eldest of six brothers who in common own the grove and, thus, the tree.

and in an intimate family rite asked her to return. Her (re)birth, including its prelude in the Goddess Place, opened ritual space to her devotees—women of the Hukumpeta neighborhood where the Nayudu family lives, devotees of her Caturugudi temple, and people who have made special vows to her. Now a different, heterogenous crowd gathers on the appointed day at the beginning of October, to carefully remove Tree-Paiditalli (for so we shall call her here) from the earth.[66] By eight o'clock in the morning, people are standing in the grove next to a tree that has been marked by turmeric powder smeared around its lower trunk, vermilion dots plastered atop the yellow, and garlands of yellow and white flowers laced around the trunk. This is Tree-Paiditalli. This time she has come not from wilderness, as she did in emerging within Big Lake, but from close to farmland, to rice paddy cultivation at a time when the rice plants are giving milk to themselves. She is a growing plant, emerging from earth, thrusting high and deep into the heavens.

A long rope is tied to the top of the tree and fastened to another to control the descent once Tree-Paiditalli has been brought out of the earth. Another tree has been cut, minimally dressed, and loaded onto a bullock-cart; this tree will become the *irusu-mānu*, the base to support Tree-Paiditalli on the Sirimanu car. Other bullock-carts are waiting to transport her, hours later, into Vizianagaram. The strong, white bullocks have been garlanded, decorated, the length of their vertebrae smeared with turmeric paste, with dots of vermilion applied on top. For the time being, these bullocks move freely in and out of the open space around Paiditalli.

Musicians play Carnatic *kritis* and devotional songs from Telugu films. Seated before the musicians on a broad reddish carpet is a distinguished-looking peasant family, the donors of Tree-Paiditalli and the *irusu-mānu*. Milling about in the open space in front of the tree are local politicians, one of the temple trustees, journalists from the local TV station and Vizianagaram newspapers, and various prominent businessmen and citizens of the town. The police too are present—initially two policemen, but soon ten and more. A kitchen has been set up to feed breakfast and lunch to the guests. The atmosphere is gentle, intimate, cozy, quiet, devotional. Bairagi Nayudu disappears to perform a puja to the local village goddess, whose consent must be given to what is to follow.

Increasing numbers of people, especially women from the surrounding villages, many holding coconuts to be broken, press forward to worship

66 Bairagi Nayudu and others often call her Ammavaru's Tree (ammavāri ceṭṭu).

18. Hugging Tree Paiditalli (David Shulman).

Paiditalli. They bring new saris and blouses which they offer to the goddess for her blessing, wrapping them around the tree or touching them to the trunk before taking them back for their own use. Some families from Vizianagaram who have Paiditalli as their personal deity (*kula-devata*) or family deity (*iṇṭi-devata*) have come to pay their respects. The ground around the base of the tree becomes a thick, muddy paste of red, yellow, and brown from all the offerings made to her. She is touched tenderly, people putting their palms on her, pressing against her. More garlands are tied around her, her upper trunk covered with flowers. By now women are pushing for openings to reach the goddess. The Vizianagaram visitors send their shouts into the air—'Victory to Paiditalli' (*Paiditallammaki jai jai*) and 'Victory to the Mother' (*jai jai Paidimamba, jai jai mata*)—their cries echoed by the crowd.

Excitement swirls and intensifies around the tree. The mood becomes one of rapt attention and subdued exhilaration. The puja activity—offering coconuts, touching the tree, circumambulating it, all at the correct astrological moment (*muhūrtam*)—becomes more frenetic. Bairagi Nayudu is standing next to the tree, his bare feet immersed in the red-yellow mud. He holds a small book of stotras for the goddess and begins to call out her epithets one by one, as the crowd responds: "Gauridevi, Bhavani, Paraśakti, Mahesvari, protect us..." The front of the tree has become a space only for men, with some women standing behind the tree. The men conduct the official rites—performing puja and felling the tree—while the women comprise the bulk of individual worshipers. At the edges of the crowd, numerous chicken heads fall to the ground, their blood daubed on her lower trunk. Yet more garments are tied around the trunk along with a golden sari with gold-embroidered *jari*. Paiditalli is becoming more alive, more restless, a goddess in her fullness, decked with *boṭṭu* forehead dots and garlands. A sense of jubilation crests through the crowd. Most conspicuous of all is the way everyone tries to touch the tree with his or her hands, placing palms flat against her trunk, holding them there, caressing her, holding tight, pushing towards her, leaning against her. Women in particular want to keep this contact going. Some grab the lower leafy branches of a neighboring tree as they lean toward the goddess. As the women embrace the goddess, their solidarity with her is palpable.

Bairagi Nayudu and his entourage circumambulate Tree-Paiditalli three times (as they would in her shrine). He takes an old iron axe with a small head, its long handle covered with turmeric and marked by two slender circles of vermilion, and makes the first cut, lightly, gently. Others of status follow, slicing lightly, tenderly. Now everyone, in a frenzy of activity, rushes to grab the axe, though at this stage only those at the top of the social and institutional hierarchy are allowed to wield it. Later anyone who wants to wield the axe lightly is given a turn, including the policemen and many of the women. This will go on for hours. The strokes are mild. Tree-Paiditalli's trunk is never seriously chopped; later her roots will be exposed and cut through. *Harati*—lustration with camphor lamps in coconut shells—is performed. The musicians have resumed playing. Now people take *bandaram*—the paste of *pasupu* and *kunkum*—from the base of the tree and apply it to their foreheads. Great quantities of *pasupu* (still used, by the way, as an anesthetic) are poured on her base, perhaps putting balm on the wound being inflicted on her (though she herself has demanded it).

The frenzied mood abates. Breakfast (*uppama*) is served as *prasadam* from the goddess. Media interviews take place. Packets of *pulihara* are given to the villagers. People say goodbye to one another, as the crowd disperses, yet buses are still on their way from Vizianagaram. Neither of the last Raja's two sons—the eldest, Ananda Gajapati Raju (as noted, still referred to as the Raja); and the younger, Ashok, the finance minister of Andhra Pradesh—has put in an appearance despite their assurances to Bairagi Nayudu, though an official-looking jeep parked nearby attests to the likely presence of an emissary.[67]

Under the direction of the Assistant Commissioner of the Religious Endowments Board, workers with long iron poles clear away the soil around Paiditalli's base, exposing her roots sunk deep within the earth. The roots are cut away from the trunk slowly and painstakingly. She is taken out of the earth, intact, with as little violence to her as possible.[68] By the late afternoon, Tree-Paiditalli is loaded onto a bullock cart and taken in the direction of the city with a brief stop for a *puja* at the Wilderness temple, and then on to Bairagi Nayudu's home in the Hukumpeta neighborhood where she will rest, perhaps to gestate until the Sirimanu, some ten days hence.[69]

---

[67] Anand told us he has never attended the Tevadam rite.

[68] The participants do not say that they are 'cutting' or 'cutting down' the tree. They say that they are 'bringing' Ammavaru. In other words, they use a language of emergence, perhaps of fullness coming into being in their world.

[69] Navaratri—the Nine Days festival in the autumn—celebrates kingship and its relationship to Durga in the little kingdoms of South India. The Sirimanu Jatra, on the Tuesday after Vijaya Dasami, the tenth day, is pervaded by ties to Durga. Paidimamba was a devotee of Durga before her death; and, as we will note, Durga will appear to Bairagi Nayudu before the Sirimanu; moreover, Paiditalli invites Durga to join her on the Sirimanu. Durga

19. Worshiping Tree Paiditalli in the Street (Boaz Amichay).

In contrast to Mud-Paiditalli, Tree-Paiditalli is neither cajoled nor cursed into becoming self-aware and so entering into the human plane. Her desire is tangible—she insists on coming, on sacrificing herself again (being cut, being injured); she demands to be taken into the city. After the more reluctant Mud-Paiditalli has grown in the Caturugudi, and after she has sown the capital city with the coming presence of kingship, she eagerly, pointedly, deliberately, reappears as a tree, her roots growing deep into the earth, her crown growing deep into the heavens. Tree-Paiditalli is the Goddess evolving further, her cosmos preparing to harvest and deliver kingship within the Old City energized and prepared by Mud-Paiditalli. Perhaps, as the sower, Pot-Paiditalli may also be growing herself as Tree-Paiditalli. Mud-Paiditalli has matured within the Caturugudi, Pot-Paiditalli returns daily to the *garbha gṛha*, but Tree-Paiditalli immediately and entirely becomes her own shrine, independent of any fixed location, first growing out of the earth but eventually becoming fully mobile as the Sirimanu, her human world entirely within herself as she acts upon it and

asks how she should come and Paiditalli suggests that she come as a bird, perched atop the Sirimanu tree and echoing what she told Appala Nayudu (Bairagi Nayudu's ancestor), that is, that she would bless Paidimamba/Paiditalli from the top of the 'flagpole', the Sirimanu tree.

shapes it. She now contrasts fully with her condition between November and May when, deep within herself, she is recursively and wholly continuous with herself, perhaps without self-awareness as we experience it. Now she is utterly self-aware, extruding and protruding into the human world within herself.

9. *Nearing the* Sirimanu: Ratham, Swing, *and* Tolellu—*the First Furrow*

Tree-Paiditalli lies quietly at the side of the street, around the corner from Bairagi Nayudu's home. Her rounded base is smoothed into a ball-like shape (*kundalaga*) likened to a pot by Bairagi Nayudu, broader than her trunk, with deep grooves where her roots had been. These grooves are said to 'open' Paiditalli. From her very top, three slender pieces, each eight to ten inches in length, are sliced away. And then, one more, the fourth, even slimmer. The longest of the three is given a crude visage, but the other two remain without identifying marks. The three are the head-body of Paiditalli and her two arms, yet no less the head-body of her younger brother, Potu Raju (the Buffalo King), more of which later. The fourth is also Paiditalli.[70] The three pieces, Paiditalli ~ Potu Raju, have a prominent positioning in events the evening prior to the Sirimanu Jatra. After the Sirimanu, one or more of the pieces (but certainly that with the visage) are placed in the tiny shrine next to the three-species tree at the back of the Caturugudi, where she ~ he joins other such presences of herself ~ himself from previous Sirimanus. These slivers retain the presence of Tree-Paiditalli after the Jatra. Similarly, the light lit when Mud-Paiditalli arrives at the Caturugudi, remaining lit until she returns the following year, keeps her presence alive there. Thus these aspects of Paiditalli never leave completely after the Sirimanu Jatra. And, as noted earlier, the three trees are the original shrine of Paiditalli when she emerged from Big Lake. Moreover, it is likely that the three-species tree is no less Paiditalli herself.[71]

Meanwhile, the full tree-length of Paiditalli lying at the side of the street is not exactly ignored, yet is hardly attended to, as she seems to gestate

[70] Sometimes these two side pieces are referred to as her two servants or her two guardians, in other words, Potu Raju.

[71] In 2009 or 2010, while the tree foliage was being trimmed, an image of Paiditalli was uncovered curving through the bark of one of them. This image was venerated at once and ever since.

horizontally in this world for the next ten days.[72] During this period there is continual bustle in the small square shaded by a giant tree in front of Bairagi Nayudu's home. Here the wooden vehicles that will participate in the Sirimanu are being assembled. Their parts have been stored in a shed that abuts the house. Now they are scattered around the yard.[73] These cars are wooden platforms on wheels, atop of which is added wooden superstructure. The most complex and delicate of these is the Sirimanu car (*ratham*) itself. Its construction has been overseen for some two decades by a draughtsman employed in the municipal department of public works. Also put together in the yard are the Anjali Ratham and the White Elephant Ratham (see below). The fourth construction in the procession is not a vehicle but is put together in the Besta neighborhood as a giant umbrella carried by these freshwater fishermen (who now are tailors). The central pole of this construction is made from the stem of a banana tree. On the morning of the Sirimanu Jatra, when the Sirimanu Ratham, the Anjali Ratham, and the Blue Elephant Ratham are pulled to the Square Temple, the Besta bring their umbrella there as well.

The Sirimanu Ratham is roughly triangular in shape, with the front point of the triangle squared off, resting on two large solid wheels of wood painted in concentric circles of yellow, green, red. The Ratham is constructed as a small platform resting over the wheel axle. Apart from its platform, the Sirimanu Ratham consists of three major parts—the vertical *irusumānu* pole, *gilika* stabilizer, and Tree-Paiditalli. The *irusumānu* is inserted into the platform to hold Tree-Paiditalli. An intricate construction of interlocking planks and poles support—under, in front of, in back of—the small platform. The wheel axle is positioned about two-thirds of the distance from the front edge of the Ratham. The back one-third of the Ratham is heavier than the front two-thirds. The *irusumānu* is shaped from the second tamarind tree taken down during the Tevadam ritual, and is replaced every year.[74] This tree is cut off much shorter than Tree-Paiditalli (perhaps one-fifth of the length of the latter). The *irusumānu*, which must support Tree-Paiditalli, is also held in place by a complex web of heavy hempen ropes.

---

[72] In 2002 her measured length was forty-four feet, with a circumference at her base of twenty-three inches.

[73] The carpenters are paid and fed by Bairagi Nayudu. He says that only his family has the secret of how to assemble the Sirimanu ratham.

[74] The term *irusumanu* means 'the vertical axle tree'. 'Verticle axle' may imply the opening of depth through revolution.

The *gilika*, stored in the shed from year to year, is fitted onto and over the rounded upper section of the *irusumānu*. The wooden *gilika* is made of four tiers held in place by two sides faced with decorative metal.[75] The first three tiers are flat planks whose narrowed rectangular ends are slotted into the sides of the *gilika*, while the top tier is shaped like a rectangular rolling pin with two round, long handles that fit through the *gilika* sides. The planks of the two lowest tiers each have a round hole through which the *gilika* is fitted onto the *irusumānu*. The third tier is solid. Tree-Paiditali will be positioned on the Ratham only right before the Sirimanu, outside the Square Temple. At that time a pin will be inserted through Tree-Paid-itali, one third of the distance from her bulbous end and two-thirds of the distance from her top, into the fourth and top tier of the *gilika*, enabling the Tree to swivel up and down against the *gilika*; while together, *gilika* and Tree-Paiditalli can rotate 360 degrees around the *irusumānu*.[76]

To the top of Tree-Paiditalli will be slotted and fastened a crude seat consisting of three pieces: a back-rest; a simple seat; and a lower foot rest. In the Jatra, once Tree-Paiditali has lifted the Priest on high at about a forty-five degree angle, her rounded, bulbous bottom is tied to the back of the Ratham. A rope from her top end drops perpendicularly to the front of the Ratham and is tied there. All the *rathams* move in the Jatra by being pulled by devotees of the Goddess.

During the days before the Sirimanu, most of the work in the yard consists of preparing Tree-Paiditali's Ratham. However, the parts of the Anjali Ratham and the Blue Elephant must also be cleaned, renewed, and fitted together. The centre piece of the Anjali Ratham is again a small platform with extended arrangements of planks, resting on two large spoked wheels. The previous year's *irusumānu* is fitted into the small platform, and onto this column is fitted a framework of poles on which riders will sit. *Anjali* (a Sanskrit word) is a two-handed *mudra* with a convex space between the hands, opening, unfolding like a bud. Yet this *mudra* is also a closed offering, mysterious, virginal—the Goddess as she has been evolving, developing into presence, coming as shapeless mud from the depths, growing out of the nourishing earth as the powerful, luxuriant shape of a tree, penetrating the depths of the sky—much like the rice shoot, in its own way.

---

[75] The term *gilika* in Telugu may refer to the stem of a tree, a stalk, or anything hanging, a pulley wheel, a toy rattle, and so forth.

[76] In February in Vizianagaram the goddess Yelamma has a small Sirimanu Ratham whose shorter tree revolves through 360 degrees.

20. Sirimanu Model in Bairagi Nayudu's Courtyard (David Shulman).

Seated atop the Anjali Ratham, we were told, will be five men dressed in saris as auspicious married women (that is, as women who have borne children and whose husbands are living).[77] As the *mudra* intimates, just as seeds bud, so do women in having children. All five are from the Pativadu family, the family of Pativadu Appalanayudu, the minister who was with Paidimamba when her brother was killed at Bobilli in 1757, who was like a father to Ammavaru when she was a little girl, and whom Bairagi Nayudu claims as an ancestor.[78] The white elephant is an open wooden cart on two spoked-wheels, its sides and trunk covered with dark blue paper, its head draped with a pink cloth (as a royal elephant might look), with place for riders to stand under the cloth. The elephant is said to signify the royal presence of the Raja. Although there are biographical elements here in Paiditalli's coming fully into presence, there is no neat linear progression in her 'biography.' Her presence is always a complex braiding of dimensions and tenses in which one strand always weaves together with others, in different formations and intensities, a cosmos folded and enfolded over and again within itself.[79]

During these days before the Sirimanu Jatra, Bairagi Nayudu undergoes a change. Usually an affable man and generous host, he becomes more closed, withdrawing into himself, resting, meditating, sleeping, preparing himself for the moments on Tree-Paiditalli when he will have, as he puts it, 'a small unconscious', yet also 'an awakening of great clarity.'[80] Around the corner, Tree-Paiditalli herself continues to lie quietly at the side of the street with the occasional devotee making an offering of fruit, turmeric, and *kunkum*, worshiping her. She rests, of course, in her great singular length, and her length is no less her great depth, pointed to explore with intention and direction the reaches of her cosmos. We can say that her

---

77 In the neighboring village of Pakki, where the Sirimanu festival takes place in January (as in many villages in the region), the five on the Anjali Ratham sit at the corners and in the middle, echoing the five directions.

78 According to another version, in the past five auspicious women sat on the Anjali Ratham, and all of them were goddesses: Kota Śakti, Mutyalamma, Nukalamma, Yelamma, Durgamma. These were later replaced by men dressed as women.

79 We were told that there also is a funereal aspect to the Anjali Ratham. The five auspicious women also recall the death of the king's sister, Paidimamba. A dead woman whose husband is still alive goes to Gauri Loka. The king's sister, who was not an auspicious woman, cannot go to Gauri Loka by herself, but she can be guided there by the five auspicious women, since she was reborn as Paiditalli, just as Sati, after throwing herself into her father's sacrificial fire, was reborn as Gauri/Parvati.

80 Bairagi Nayudu says that on the Sirimanu he is not conscious (*nāku calanam uṇḍadu*) in the sense of not being interactive, not paying attention, not responding, having no awareness of self and other during this time.

maturation and intensification lie in this form. First an amorphous amalgam of earth and water—mud—when she first surfaced onto the shore, then a fertile catalyst for the *uru*, she is now shaping and honing herself to bring king to kingdom.

Some days later, on the Sunday eve[81] before the Sirimanu on the Tuesday, after a long day of laboring on the *rathams*, the swing (*uyyala/ wiyyala*) is set up in front of the Square Temple, the work ending close to midnight. Swinging is a still-mysterious dynamic in temple-related activities in South India. Many large temples (and larger private homes) have permanent swing installations. Young women of puberty age swing (seeking husbands), young couples swing before their marriage (seeking fertility), deities who are processed outside of their temples sometimes swing (vibrating with energy and force) in their *rathams*, devotees swing on hooks inserted beneath their back muscles in devotion to their deities (notably, Murugan), and so on.[82] We heard no exegesis about swinging or what it does to swingers and others, apart from being told that in the case of deities this activity pleases them. Our sense is that swinging is indeed a dynamic, one related to interior movement and conditions of being, and perhaps to temporal movement (Chapter Two). However, the dynamic is not uni-directional. Sometimes the arcing, to and fro, seems to activate, focus, concentrate, and intensify interior dynamics. Yet sometimes this arcing, or just sitting on the swing, seems to activate dynamics that quiet, soothe, and perhaps reduce the intensity of interiority, as we shall see.

The swing will remain in place until Paiditalli, on the Tuesday two weeks after the Sirimanu, is 'swung off' to the Wilderness Temple until the following May. But on that Sunday evening before the Sirimanu, as the worship of Tree-Paiditalli begins in earnest around the corner from Bairagi Nayudu's home, Paiditalli of the Square Temple will emerge, as will the pots that are Paiditalli, to sit on the swing together with Bairagi Nayudu and his younger son, Venkat. The seat of the swing is hung on chains attached to a metal crossbar supported by four angled, metal poles. All the metal parts are painted red. The seat itself, broad enough for a few people side by side, is painted yellow, with a large lotus in its middle.

---

[81] In 2002 this Sunday is the eve of Vijayadasami. As noted, the Sirimanu takes its chronological lead from the celebration of Durga.

[82] In the village of Pakki, near Vizianagaram, the Swinging Away of the Goddess (*uyyala-kambālu*) has strong erotic components, as it does, apparently, for people who swing there. According to one informant in Pakki: 'Everyone swings in the swing. They perch their asses in the swing and move a little [swing] and fuck off.'

Significantly, the swing as a vehicle resonates powerfully with the Sirimanu Ratham. The metal poles replaced wooden ones long ago, but two of these wooden poles are still fixed in place on either side of the seat each year. The poles were once part of Paiditalli's Sirimanu Ratham. Strung between these wooden poles along the metal crosspiece from which the seat is hung is a hempen rope that was once part of the rope used to pull Tree-Paiditalli's Ratham during the Sirimanu. The swing is another version of Tree-Paiditalli's vehicle.[83] So, too, in two weeks time the swing-*ratham* will swing Paiditalli off to the Wilderness Temple.

The seat is bathed, and a *yantra* design is drawn on the ground under it. Bairagi Nayudu consecrates the seat as he would an icon of the Goddess within the *garbha gṛiha* or, for that matter, the *garbha gṛiha* itself. He washes himself and the seat of the swing, offers it turmeric, *kunkum*, and *arati*, and re-enters the temple. The two clay pots of Pot-Paiditalli emerge, the first carried by Paidi Raju, the low-caste servant of the Goddess who went into the lake to await her coming and who carried her in the *jangidi* from there to the Square Temple. The clay pots are followed by Bairagi Nayudu, holding a small, golden mobile image, *utsava mūrti*, of Paiditalli, accompanied by Venkat. The *utsava mūrti* is placed on the lotus design, Bairagi Nayudu sitting to her right, Venkat to her left. Then Bairagi Nayudu sits on the lotus, holding Paiditalli, with Venkat next to him and a clay pot to either side of the pair. Together they swing gently, peacefully, for a time. The *utsava murti* returns inside the Temple, into the *garbha gṛiha*. There the permanent image, the *mūla mūrti*, of Paiditalli is being washed, rubbed with turmeric, sandalwood, milk, and coconut milk. By 2 AM, she has been dried, dressed in a red sari, and draped with flowers, a silver lion to either side and her *utsava mūrti*, also in a red sari, in front of her. In the meantime everything else had been removed from the *garbha gṛiha*, which was washed and cleaned.

The following evening, Monday, the Goddess pots will make a grand visit to the Kota Śakti, and on their return to the Caturugudi they will come together with the three-piece Paiditalli ~ Potu Raju, before the Sirimanu Jatra the following day, Tuesday. During the two weeks between swingings, all of the aspects and qualities of Paiditalli come together (and also fall apart or separate once more) in different combinations. The swing is a *modulator* of the interior intensities of the Goddess, helping to regulate, perhaps to focus the self-organization of her intentionality and

[83] Later on, women praying for fertility will place new saris across the rope.

feeling. This is especially so in relation to the Goddess pots, even though ultimately the pots cannot be separated from other foci of her presence, Tree-Paiditalli, the three-piece Paiditalli ~ Potu Raju, and the temple *mūrtis*. This modulation becomes especially significant as cosmos begins to galvanize for the climactic harvesting of kingship after the triumphal re-growth of kingdom, thus overcoming the catastrophe of Bobbili. As we noted, swinging may augment, intensify, excite, yet no less relax, soothe, quiet. The capacity for excitation may be more self-evident (witness the qualities of the Sirimanu Ratham built into the swing), yet the potential for soothing is also there, if more implicitly. Swinging the Goddess is also a kind of putting her to bed, to rest, to sleep (*pavaḷimpu seva*).[84] We note that the swing is also thought of as a cradle. Years before, in a practice no longer done, Harijans (Dalits) covered the swing seat with the skins of goats and roosters they had sacrificed to the Goddess. The sacrifices were intended to satisfy the Goddess, and so to soothe and quiet her excitability and anger. Within her sacrifices, one may say, she could rest within her cradle, rest deep within herself.[85] The swing is erected and first used as her presence intensifies and excites, beginning with the Sunday. Yet when she finally swings away to the Vanamgudi, she is quiet, relaxed.

Beginning on that Sunday evening, Tree-Paiditalli awakens, her self-intensity growing. Her entire length, her body, is rubbed with turmeric; red rings of *kunkum* are traced in threes at different points around her circumference; and a lotus of *kunkum* is drawn near her *bulbous* base. Camphor lamps are placed along her entire length, the wicks lit. Devotees, mainly women, begin to come more frequently to her rounded base with offerings and flowers, while coconuts are broken there. During Monday, and especially Monday night, these activities quicken, as does the presence of Tree-Paiditalli.

Turned yellow with turmeric, red with vermillion (just as the inside of her Tamarind is yellowish and reddish), decorated and adorned, looked at from rounded bottom to tapering top as she lies in the street, Tree-Paiditalli in reverse provocatively begins to look something like a huge snake (*nāga*) with a thickened, blunt head. Remember that when grooves were smoothed into this head-like form (emphasizing where her roots had been) after she first came to lie in the street, these grooves were said

---

84 This is the term used in the Vanamgudi, when she rests, or goes elsewhere into herself, for half the year.

85 Put otherwise, her cradle is already enveloped in wilderness, in the skins of the sacrifices.

to 'open her'. Now they resemble a mouth. These thoughts may seem far-fetched and without basis in local exegesis, yet perhaps they should not be dismissed out-of-hand. The nagas dwelling beneath the ground epitomize self-regeneration as they periodically shed their skins in revival.[86] Paiditalli joins earth (the mud of the lake) and the heavens (into which she thrusts during the Sirimanu itself). Just as she is said to be able to move in all directions on the Sirimanu Ratham, so her forming here enlivens the trajectory of earth and sky. Her forming is malleable: in one direction she opens to the nether world, in the other, into the heavens, joining both (or both emerging) through her being. She grows into the earth no less than she grows into the sky. Put this way, and despite her material appearance of utter lineality, her dynamics are closer to a recursive arcing, resonating with the relationships of waters and mountains—of depth in all directions—that we discussed in the first section. In other words, she is becoming the continuous self-enclosing cosmos that now includes the furrowed, sown human world within which king will be brought into kingship and kingdom.

During Monday the final work on the *rathams* is completed, and the intensification of Tree-Paiditalli through offering (and rooster sacrifice) continues. Bairagi Nayudu fasts from dawn on Monday until he descends from the Sirimanu at dusk on Tuesday. On Monday morning, Anand, the Raja-who-would-have-been, offers a new sari, *pasupu*, and *kunkum* to Paiditalli in the Square Temple. When we see him later that morning (in 2002) in his office in the Kota, he seems rejuvenated from his meeting with her. On Monday evening the mood becomes feverish; the tempo and fervor of offering and sacrifice increase greatly. Paiditalli is caressed over and over; she is enveloped in intense prayers as numerous coconuts are broken around her, and chicken heads are lopped, their blood smeared along her length.[87] Monday night is the time of the Tolellu, which translates literally as 'first furrow'. The term, 'first furrow', is highly suggestive, joining once more agriculture and sexual generation, though in the case of Tolellu these dynamics are oriented to the future.

Tolellu is the last major movement of the Goddess, that is, of Pot-Paiditalli to the Kota before the Sirimanu. The visit is described as inviting

---

[86] The bards who will sing the story of Adi Śakti at the Round Mahal and who sometimes sing Paiditalli's praises at the Swinging Away say that she is connected to snakes. Keep in mind, too, that when the farmer refused to surrender the selected tamarind tree from his grove, his wife was threatened in a dream by a snake.

[87] In South India the coconut, with its shape, hair, and (dimpled) eyes is often likened to a human head, and its breaking to a substitute for human sacrifice.

Kanaka Durga to the Sirimanu, again intimating the close relationship between Durga and the Kota Sakti. Yet the visit is also the culmination of Pot-Paiditalli's contacts with the Kota Śakti. Indeed, it marks the presence at the Round Mahal of all of Paiditalli that has fully evolved thus far, distributed and refracted through the variety of pots that journey there the evening before she, Tree-Paiditalli, brings king to Kota.

On the Monday the pots are placed in the *garbha gṛiha* of the Square Temple, next to the *mūla mūrti* and *utsava mūrti.* Outside, worshipers, especially women, throw packets of *pasupu* and *kunkum*, offerings to please the Goddess, at the façade of the Temple. After dark the pots, decorated with *kunkum* and garlanded are assembled inside the Temple. As Bairagi Nayudu walks through the Temple, parents gently place their infant children beneath his bare feet, so that he passes over them. This is thought beneficial for their good fortune. Somehow, with excellent coordination, he does step over them without stepping on them as he makes his way. The pots emerge from the temple and sit on the swing in various combinations. At around 11 PM they are put on the heads of carriers. Those pots with flat tops are covered with a plate of camphor with a lit wick. In all, ten pots (the two clay pots, the two five-metal pots, and the six brass pots) are taken in procession. All are carried by family right, the two clay pots by women. Saris are spread before the pot-carriers so that during the hundreds of meters to the Kota the Goddess will never touch the ground. The atmosphere is frenetic, bodies packed closely together around the pot-carriers. Every body, everybody, is participating in this sending forth of the goddess to her meeting with the protectress of palace kingship—by pressing together, surging forward, putting forth hands to hold the pots in place, to balance the pot-carriers, to touch the goddess-pots, by clutching at and supporting one another as an entirety, a massive presence.

The pots stop at particular private homes on their way to the Kota and then on their way back to the Square Temple. Thus they come to the homes of the descendants of two of the three ministers who ran the kingdom after the king was killed during the Bobilli war—one home is visited on the way to the Kota and the other on the way back. During the home visit on the way to the Kota, Bairagi Nayudu, dressed in white, garlanded, a silk scarf around his neck, receives the Maharaja's turban of white with a light blue border that he will wear during the Jatra.[88] He puts it on, leaves it in the home, and receives it again later on. The silk clothes he will wear

[88] Until over twenty years ago the turban (*talapaga*) consisted of six yards of cotton material that only one man in Vizianagaram, a *Burrakatha* bard, knew how to tie. Since

21. Tolellu: Five-Metal Pot on its Way to the Fort (David Shulman).

22. Pots Moving to the Fort, Tolellu (David Shulman).

are stored, he says, in the Raja's palace.[89] Bairagi Nayudu knows full well that when riding Tree-Paiditalli he will be dressed as the Raja, adding that, 'The first time [his ancestor] Appala Nayudu was about to climb onto the Sirimanu, the Raja put those clothes on him'. He refers to the clothing as 'royal attire'.

The procession enters the Kota through its great gates that front on the thoroughfare along which Tree-Paiditalli will travel the following day. To the right of the entry port, high on the wall, is an arched niche. As the procession approaches, Bairagi Nayudu sees Kanaka Durga standing in the niche, protecting the walls of the palace, her presence unwavering. The procession of pots, augmented by hundreds of devotees and spectators, winds its way within to the Moti Mahal, the Round Mahal (now an educational institution for women) and stops before the entrance to the building which is shut, locked, dark. The area too is dark and compacted

---

his death, the turban has been brought from Rajasthan, of a kind used in weddings, simply worn, like a cap.

89 It is not clear whether he is referring to storage in the Kota complex or in one of Anand's homes or his office. Anand, in one of the very few acts that in practice connect him to the Sirimanu, said that he gives the clothes to Bairagi Nayudu on the Monday morning, when he goes to worship Paiditalli in the Square Temple.

with people. The mood is solemn, quiet, attentive for this climactic meeting of Paiditalli and Kota Śakti. Paiditalli is introduced by the bards who sing her epic of origin. Here she becomes the primordial śakti, Adi Śakti, who created the cosmos, attesting to her common origins with the Kota Śakti and Kanaka Durga.[90] But the bards also sing songs of sadness (*dolak*), lamentations of mourning, melancholic, full of distress, beauty, affection, dirges fore-grounding Paidimamba's death as the sister of the king.[91] There is here an implicit condensation of her cosmic biography that on the morrow, through the Sirimanu, will become her triumph. At the human center of this recitation is Bairagi Nayudu. He sits facing the Moti Mahal, enfolding his teenage son, Venkat, with both arms. If all goes well, the Kota Śakti manifests herself in close proximity to Paiditalli by entering within and filling Venkat with her presence. He bows his head, holding his face with one hand.[92] After this, the pots make their way back to the Square Temple. On the return journey the mood is more relaxed, the crowds looser, more dispersed.

The coming rites of Tolellu, the First Furrow, are the rapid coming to maturity of Tree-Paiditalli—the coming to the apex of her intentionality and awareness in the human portion of her cosmos. The rapidity of this dynamic shifts directly into the accelerated rhythm of her climactic

---

[90] For another Andhra ritual context in which a local goddess is identified with Adi Śakti, see Chapter 3; Anand n.d.

[91] In 2003 there were heavy rains before and during Tolellu. On the night of Tolellu the pots were late in leaving for the Kota. The priests felt there was an obstacle of some sort. A priest from the Wilderness Temple (Bairagi Nayudu's son-in-law) was possessed by Paiditalli who told him she wanted to hear Dolak songs even before the procession began. But perhaps because of the rains, the bards did not come. At the Round Mahal the water was above peoples' ankles. The story of Adi Śakti was not sung; neither were songs of Dolak; and Venkat was not possessed by the Kota Śakti.

[92] The Kota Śakti enters Venkat (one can say that Venkat enters the cosmos of the Kota Śakti) because his father intended that Venkat succeed him as Paiditalli's priest and so, Venkat would ride the Sirimanu. Riding the Sirimanu, Bairagi Nayudu has kingship within him—he is Paidimamba's brother, the king. Next to the Moti Mahal, Bairagi Nayudu deputes the presence of kingship to Venkat, holding him in his arms, so that the Kota Sakti once more receives the presence of the king. However, Venkat did not succeed his father. In the past the decision on succession would have been made by the Raja. Now it was made by the Board of the Caturugudi, and the appointment went to the husband of Nayudu's eldest daughter, Nirmala. Her husband was a priest at the Vanamgudi, the Wilderness Temple. However Nirmala herself said that Ammavaru can never be taken out of her priest (by a bureaucratic decision) and that those who replaced Nayudu, including her husband, have become partially paralyzed in their limbs (which they try to conceal) because of their inability to handle the intimate presence of Ammavaru. So, Nayudu also has the Goddess within him on a permanent basis, and as such he has some presence in her depth as well as on her surface.

activity in the Sirimanu Jatra itself. The area in front of and around the Square Temple is very crowded, shoulder to shoulder, surging into dense movement, floundering on obstacles, waves breaking, and again into movement, the air heavy with wood smoke, incense, spice. Already there are long lines waiting to enter the Temple for *darshan* of the Goddess.[93] With the exception of the two clay pots that are put onto the swing, the pots are placed on a stepped *saduru*-platform (used as a *balla* or table, says Bairagi Nayudu) just across the street from the entrance to the Square Temple. The three-piece cut from the Sirimanu tree that we are calling Paiditalli ~ Potu Raju (with the interior fluidity of its coalescences as one and the other) makes its first public appearance. Bairagi Nayudu wraps the three pieces together in red cloth and places them in a basket together with five bamboo sticks, paddy, and basic peasant fare (an onion, *kūḍlu*) to feed Paiditalli~Potu Raju inside the basket. This is resonant with the re-birth and nurturance of Paiditalli emerging from Big Lake. What may be involved in Tolellu is the birth and nurturance of the Potu Raju qualities of Paiditalli which emerge here for the first time in the Paiditalli ritual cycle. Nayudu then takes out the three-piece, removes the red cloth and wraps her~him in black cloth, and then in white cloth, placing her~him in the white cloth between the two five-metal pots (each of which is now topped by five lit wicks) on the *saduru*-platform.[94] Tree-Paiditalli has joined Pot-Paiditalli (who of course is also Mud-Paiditalli). Together the Paiditallis watch the performance of *Asarlu*. However here Potu Raju should be introduced a bit more thoroughly.

Potu Raju (Buffalo King) is the generic younger brother of the Goddess in South India. Where the Goddess is present, his presence is ubiquitous. He is often a small slab of rock stuck in the ground or an equivalent wooden post, stake, or spike. Often he stands as such outside and opposite a goddess shrine. He is considered the guardian and protector of the Goddess. Yet consider his name in relation to Mahisasura, the great Buffalo Demon killed by Kanaka Durga at the end of Navaratri, and Durga receiving buffalo sacrifices during Navaratri.

---

93 For the first time in this cycle entry to the temple is by payment. There are two lines, an express line (costing more) and a regular one.

94 The color red, we were told, is that of life, of intense feeling (*raudram*). That of black is death, sadness, sorrow, mourning, and fierceness. White is the quality of purity (*svaccamayinadi*). Here again are condensed the histories of the princess, Paidimamba and her brother the king—their life (red), his death, her mourning, and her death (black), her return as the virgin goddess who yet contains her (now) younger, less senior brother (white).

Bairagi Nayudu identifies the middle piece with the crude visage as Paiditalli, saying it is carved in her shape, and the two side pieces as Potu Raju, her guardian. Identity is crucial here to our argument, or, more ambiguously, the infra-lapping of identity. When we asked a devotee of Paiditalli, one who participates in many of her rituals inside the Cadurugudi, about the identity of the three-piece, he responded immediately, unthinkingly, that the three are Potu Raju. Then he denied this and insisted they were Ammavaru in different forms. An interesting parallel comes from southern Andhra, courtesy of our friend, Joyce Flueckiger (personal communication July 2010), who has done intensive fieldwork on the goddess Gangamma in Tirupati. A central figure of wood in her Jatra is one that Handelman (Chapter Three) identified as Gangamma. Years later, one of Joyce's main informants (himself a Gangamma ritualist) who earlier had said that this figure was Gangamma now said that this was Potu Raju. The next day in Gangamma's shrine, looking at this figure, this man said, 'He's *śaktiswarupani*.' In other words, he is, is part of, the Ammavaru who is called śaktiswarupani, the creatrix in her 'base' (*mūlam*) or elementary form.

Ethnographically, these ambiguities are quite on the mark in Paiditalli's cosmos. The three slivers are cut from the top of the Sirimanu tree who is Paiditalli. The three pieces come out of Paiditalli.[95] Potu Raju comes out of Paiditalli. In a sense she creates, gives birth to, her younger brother, just as everything emerges from her in her cosmos. She and he share elemental qualities of being, even as she may transform him from Buffalo Demon (her great enemy) into Buffalo King (her guardian). The Paiditalli~Potu Raju relationship is deep, and depth in this cosmos is a fluid intensity. Therefore, male and female move through one another, *inhabit* one another, are never completely separated from one another. Paiditalli~Potu Raju are one and are two. They are fluid coalescences within Paiditalli, into one vector, into another. Potu Raju appears in this way just prior to and during the Sirimanu because of his powerful resonance with the king, the older brother of Paidimamba who became Paiditalli with her own (now) younger brother, Potu Raju. In both instances the bond between sister and brother is profound and intertwined with death and resurrection. And all four of them ride the Sirimanu as kingship is harvested and brought to the center of kingdom.

---

[95] We know from the work of Madeleine Biardeau (2004) on Potu Raju in South India that his wooden post or stake is almost always cut from the Śami, a form of pea tree. Yet in Vizianagaram, the Potu Raju of Tolellu and the Sirimanu is always cut from the Sirimanu tree, from Paiditalli. Nonetheless, in South India the Śami tree is considered female.

Facing the Goddess pots, Bairagi Nayudu, Venkat, Paidi Raju, and two others do *Asarlu.* Each of the five takes up a bamboo stick from the basket. Each stick has a cotton bud at its tip. The five light the buds and, moving together as a line, walk toward the pots and Paiditalli~Potu Raju and back three times, holding the sticks and clapping their hands. Then they dash the sticks to the ground, extinguishing their flaming buds. *Asarlu* is a standard segment in other *Tolellu* rites held elsewhere, especially in villages during January—February. During *Asarlu* in these settings, the story of Śiva killing his demonic son Andhaka is told or referred to, the demonic drops of blood (perhaps here the flames?) spilling on to the earth, giving birth to other demons, until the blood is caught in a vessel by the goddess Camunda or by Śiva himself (see Handelman and Shulman 1997). *Asarlu* appears as an anti-demonic rite, cleansing the earth of demonic fertility at the place where big Tree-Paiditalli will mount her Ratham in some hours' time.[96] No less, here *Asarlu* appears to protect Potu Raju, newly emerged from Paiditalli.

In this space there immediately follows *Sirilu,* the First Furrow itself, whose name connotes abundance, wealth. A new, white, cotton sari is placed on a wooden platform, opened, a basket filled with rice seed (*dhanyam*) placed within, the cloth then curved, furled over itself, folded. Two men dressed in saris (in other words, dressed as women) kneel on the platform, one at the head of the sari, the other at its foot.[97] In the lap of one is the three-piece, still covered in white cloth.[98] A large white cloth covers the men and the sari so that their movements are partially cloaked. With great speed the platform is lifted at outstretched arm's length high in the air and whirled clockwise quickly three times. Beneath the covering cloth the two men-women rapidly twist, turn, and churn the ends of the furled sari with its basket of paddy back and forth, over and again. First Furrow—this twisting and churning of the folded sari within the twisting and turning of the cloth-concealed platform—takes no longer than a couple of minutes. Bairagi Nayudu clambers up onto the *saduru*-platform, briefly tells the story of Paiditalli, focusing on her desire that these rice seeds be distributed to the agriculturalists, and then distributes the seeds

96 Here Śiva's killing of Andhaka resonates with Durga's killing of Mahishasura, to which the Sirimanu Jatra is calibrated.

97 The two men wear a type of sari called *veyyigalla cheera* (a thousand squares or a thousand eyes) or *banala cheera* (a sari of arrows). Both are of inexpensive cotton and dear to Ammavaru. One of the men is usually Bairagi Nayudu, but he was ill and deputed Venkat to take his place. The other was the priest in charge of Tolellu.

98 Interview with Bairagi Nayudu in his home, 29 May 2009, 10 AM.

to farmers (*raituluki*). The farmers will mix these seeds with those in their storage spaces so that the former, energized, fertilized, will do the same to the stored seeds.

The three-piece, Paiditalli~Potu Raju, is treated by Bairagi Nayudu as an infant, gently put into a basket—into womb-like space—together with nourishment (as was little Mud-Paiditalli) but now also with weapons, the *bamboo* sticks that will be used to rid this space of demonic presence. Little Mud-Paiditalli, the very beginning of the entire Goddess dynamic, was entirely a female infant, while the three-piece is in a sense the dynamic of Paiditalli giving birth to her younger brother. The sticks protect but likely are also empowered even by this infant. Similarly, the aspects of Paiditalli unite at *Asarlu*—and while certain of the aspects (the five-metal pots) will return with her to Vanamgudi and others (most of the other pots) will enter the quietude of storage, the presence of the three-piece will endure in the tiny shrine nestled against the three-species tree.

The First Furrow, this hidden, secretive churning of the paddy-filled basket-womb in the white sari, suggests generative force, sexual in its intimations. Producing seed by churning is called *manthana* in Sanskrit and Telugu—the vigorous back-and-forth movement, *pravritti* and *nivritti*, as is the case with butter churning, with the churning of the mythic Ocean of Milk, and with sexual churning; also with the generation of fire by rubbing together wooden fire-sticks in a rotating, contrary manner, the 'male' stick inserted vertically into a hole in the horizontal 'female' stick and rotated rhythmically back and forth by means of a rope, generating smoke and fire. Fire is thought to be a potentiality of the wood, latent in the wood all along. The logic of churning seems close to that of the churning of paddy in the Tolellu sari—the sari that can only be the goddess herself, virginal and bursting with potency barely contained. First Furrow brings the Goddess to puberty, reaching maturity, awakening her sakti into its highest intensity. This is how she will appear on the morrow—as a virgin Goddess endowed with the creative energy and fertility of *śakti*, in a sense generating kingship in a virgin birth.[99]

[99] On the *Sirilu* platform are two female-males (the priests), one shifting female~male, and the virgin Goddess. The rite is female and female-centered, yet with a male presence. One hint of what may be occurring there evokes another variety of sexual potency and fertility, which is that of drawing this force out of the female. This hint comes from a secret rite in the Jagannatha temple in the former little kingdom of Puri in Northern Kalinga. This undoubtedly tantric rite is discussed by Apffel-Marglin and Hudson (2008: 79–84) and we will not enter into it here. Yet, significant for our thinking is that the purpose of the Puri rite is to draw sexual fluid from the devadasi (one of a number of temple courtesans

Paddy is a potentiality of the Goddess in her cosmos; she activates this seed with a view to the future, to her next coming to the kingdom. Just as she has self-generated herself and sown the space of the city, so she sows time, future time, when she will return to harvest both kingship and the prime crop upon which the kingdom most depends. The First Furrow, then, extends or projects the current cycle before it culminates and closes into the onset of the next. More accurately, First Furrow is an extension of futurity, creating depth, opening space/time. The rite shows depth in its very making—the height of the platform, the textured folding of the sari, the layered covering of sari and women-men, the whirling of the platform as the sari is folded and twisted and turned. The very name of this rite accentuates the potential for growth latent in the opening of depth by sowing. First Furrow is perhaps an originary cut, a primordial opening that enables human being to reach towards the interiority of the goddess. Perhaps we can say that with First Furrow the coming yearly cycle is already a curled foetus folded into Paiditalli's cosmos.

During the activities of the Tolellu evening and night, there is a tremendous arousal and intensification of the energy of Śakti, first through the meeting of Pot-Paiditalli and Kota Śakti and then through the emergence of Potu Raju from his older sister, Paiditalli. First Furrow is a sort of assurance on the part of the Goddess that she will return during next year's month of Vaisakha. Harvesting (the Sirimanu morrow) and awakening and sowing (the First Furrow of today) are not separated lineally, in neat chronological sequence. Next year's sowing of the kingdom by the fully evolved and wholly present Goddess precedes her harvest of this year. We underscore once more that the temporality of this dynamic is fractal—next year and this year emerging from within one another, Moebius-like, next year and this year slipping, gliding through one another, fructifying one another. These too are the interior dynamics of the goddess: she and

married to the deity, Jagannatha, and therefore forever virgins regardless of the sexual services required of them in the temple). The sexual fluid (*rajas*) is essentially the śakti of the female and of course the goddess; and this sexual fluid can release *tejas*, the brilliant conquering power of the goddess which protects the king, the further purpose of the rite. Now, we are not saying that the Vizianagaram *Sirilu* is comparable to this Orissan Puri rite. Yet there are dynamics in the two rites that resonate. The purpose of the *Sirilu* is that Paiditalli fertilize the rice seeds in her 'womb', and she must do this as the virgin goddess. This may well involve bringing forth, or in this case deeply involuting the sexual fluid that is her śakti. Turning, twisting, and churning her may indeed be involved in this, her śakti/sexual fluid activated, entering the rice seeds. Perhaps she does for the rice what she does for Potu Raju, interiorizing the former, exteriorizing the latter.

the kingdom are inseparable, moving and moved by one another, resonating powerfully together.

The activities of Monday night compact the biography and dynamics of Paiditalli. The epic of Adi Śakti tells of the creation of the Goddess and her creation of cosmos. The songs of *Dolak*, their sadness and mourning, imply her death—the death of Paidimamba, the king's sister. *Asarlu* cleanses the grounds on which the regenerative power of the Goddess will shortly begin to operate and from which her harvesting of kingship will depart. *Tolellu* is the maturation, intensification, and activation of this generative power toward its own future Then, on the morrow, the Sirimanu marks the harvesting of this year.

Schematically, these movements may appear flat and linear, but they are not so in a cosmos that curls within itself, folding and unfolding. Consider that 'pregnancy' both of rice and woman is a folding into itself, or an unfolding from within itself, and that rice is self-pollinating and the goddess, self-generating. Thus Paiditalli folds into herself as she journeys back to the Wilderness Temple after the Sirimanu. The following Vaisakha she unfolds into the human world as mud that in itself is folded into itself within water, sticking to itself. Mud then unfolds into the clay *ghatalu*, as the *ghatalu* unfold into the city, even as Paiditalli folds into herself as the Tamarind tree, later to unfold into the space/time of kingship in the Sirimanu as her priest folds into kingship that unfolds into its kingdom. Interior is turning into exterior that is turning into interior . . . Throughout these dynamics, the human world is within Paiditalli, even as she appears within this world.

10. *The Sirimanu Jatra*

This Monday evening is the full moon night of the Telugu month of Asviyuja. Tuesday daybreak in 2002 is cloudy, with the possibility of rain. By early morning there are already three long lines of thousands of people waiting for *darshan* of the goddess in the Square Temple. The line stretches around the block, hundreds of yards down the thoroughfare the Sirimanu will travel, almost to the Kota. The city is full, bursting with visitors and vigor. Rough estimates in 2002 and 2003 were that some 300,000 to 400,000 persons came to see the Sirimanu Jatra. In the Besta neighborhood final touches are put on their giant umbrella made of fish netting stretched over a frame with cut-outs of fishes attached to the net, topped with a large gold and silver-covered cardboard fish. Besta say that the umbrella has a thousand eyes, a reference to village disease-bearing goddesses who are said to have a thousand 'eyes', that is, the pox sores on the bodies of their victims. In Besta eyes, Paiditalli is a disease-bearing and disease-healing goddess. The young men who will carry the umbrella have yellow conical hats tied under their chins; they look like they are carrying halved bananas on their heads.

Tree-Paiditalli's length is again rubbed with *pasupu* and balanced length-wise on an ox-cart; the oxen have painted horns, garlands around their necks. People hug her, throw water on her, break coconuts. From Bairagi Nayudu's home the bullocks pull her to Four Lantern Street in front of the Square Temple. Behind her is pulled the Sirimanu Ratham. Bairagi Nayudu, composed, reserved, but shouting orders like a Raja, stands on the Sirimanu Ratham, with young men hanging from it in all directions. Behind the Sirimanu Ratham is pulled the Anjali Ratham with some twenty-five men on it, roaring obscenities at those watching from the sides. Behind the Anjali Ratham comes the white elephant (colored purple), also packed with men. All move slowly through the crowded streets. Opposite the Square Temple, Tree-Paiditalli is shifted with care to the Sirimanu Ratham, pinned to the *gilika*, tied in place, the seat fastened to her upper end, a bunch of bananas tied around her just below the seat.[100]

The balconies, rooftops, and windows along the route fill with people; the streets are dense with movement. The Besta arrive with their multicolored, beribboned, fish-decorated umbrella on its banana tree stem; so too does the 'flow of milk' (*pāladāra*), the 'army' of Paiditalli organized by Bairagi Nayudu—young men who have taken a vow to serve her and who in this regard are observing a forty-one day vow of abstinence. The 'soldiers' wear red bandanas around their necks or foreheads and carry staves that they crack against one another's. The police armed with long bamboo lathis are evident in force, led by their (female) district superintendent (who has a Ph.D in Sociology). The police begin stringing rope along both sides of the route. Later they will work at keeping spectators behind these barriers, liberally threatening blows with their batons to do so. They will also surround Bairagi Nayudu when he sits in the wooden seat attached to Paiditalli and is lifted high in the air; similarly, when he dismounts.

100 In 2002 the beginning of the Sirimanu was delayed by some forty minutes until the police were satisfied that the pinning of Tree-Paiditalli to the Gilika was carefully accomplished, to ensure safety during the Jatra.

23. Transporting Tree Paiditalli to Square Temple (David Shulman).

Slowly, amidst jostling, calls, cries, shouts, the groups and *rathams* assemble at the head of the Sirimanu route. The Bestas are fighting amongst themselves, pushing, shoving, some punching until the police pull them apart. The long line of worshipers waiting to enter the Square Temple for *darshan* of the Goddess continues to move steadily, step by step. The Raja-who-would-have-been or his brother are expected to take a pull or two of the Sirimanu Ratham rope, then to sit further down the route on the wall of the Kota, opposite the tomb of the Muslim pir.[101] Neither are to be seen at the head of the route, though Anand told us the previous morning that he would be present. Perhaps he sits on the Kota wall.

Bairagi Nayudu appears from within the Square Temple and sits next to Tree-Paiditalli. He is barefoot and is without his customary eye-glasses. He wears the white silken finery of a raja, with the raja's high, conical, Rajasthani turban on his head. He is garlanded like a groom and receives *pasupu* and *kunkum*. He levers himself into the wooden seat and is tied in

[101] They are expected to sit next to a sign of royalty, a flag that flies on the wall of the Kota. It is said that the flag was given to a Vizianagaram king by a Muslim Maharaja of Hyderabad as a sign of gratitude after the former saved the abducted son of the latter. The design of the flag is that of two swords with curved blades angled towards each other, their tips meeting.

24. Bairagi Nayudu before Mounting the Sirimanu (Boaz Amichay).

place with new saris, his feet resting on the low wooden footrest (Later, as he enters the goddess, his feet will dangle in the air). In one fist he holds tightly to a stick to which is attached a small, squarish mat of rattan. The stick is the fourth sliver cut from the Sirimanu, Ammavaru Paiditalli, and the rattan is the square mat on which the Goddess used to sit under the three-species tree, her originary shrine after her primordial emergence as Paiditalli from Big Lake. Bairagi Nayudu holds tightly to Paiditalli throughout his ride, even as she carries him. In his lap, wrapped in a silk cloth, is the three-piece, Paiditalli~Potu Raju, the other three pieces cut from the Sirimanu tree.[102] Water is thrown liberally over the *bulbous* root end of Tree-Paiditalli and onto the Besta umbrella covered with fish-nets and fish designs.[103] Perhaps the most chaotic, emotional, yet apprehensive moment in the entire Sirimanu is when Bairagi Nayudu is tied into the seat with new saris. We return to this below.

102 Interview with Bairagi Nayudu in his home, Friday, 29 May 2009.

103 Again, the Besta umbrella resonates with a mountain within the sea (of fishes and fish-nets), a mountain of holes filling within with water, a mountain turning into a wave.

With a great cry that sounds like a wave-like sigh, Tree-Paiditalli lifts her priest high in the air at about a 45 degree angle and (in the relatively open street space close to the Square Temple) swings him in an arc of some 180 degrees, showing that she moves in all directions. Her *bottom* with its rounded *tumbu* is then tied in place on the Sirimanu Ratham, and she does not move again sideways. In our understanding this great raising and heightening of space is the opening of the *depth* of the kingdom by Tree-Paiditalli, harvesting its capacity for creativity and growth as rejuvenated by the goddess. Depth, we argued, is watery space in which heights fill within with water and fit into or submerge within watery depths. Watery space in this cosmos is crucial to growth; and the task of the Goddess is to turn the dry, dusty, flatland of civilization into the depths of watery space. Paiditalli emerges annually from the deep fluidity that is her existence, as does a rejuvenated kingship, synchronized with the growing of rice, that will be brought to its rightful location, the palace-fort, on the head of the goddess. As she moves through Vizianagaram she distributes blessings to everyone.[104] Put more forcefully, king and kingship *sprout* from within Paiditalli into her priest, the receptacle formed to receive them within the human world within her cosmos.

In 2002 and 2003, at the very forefront of the Jatra is a Fisherman who begins its movement. He appears in trance, supported on both sides, leaning backwards and sideways as if he had limited control over himself, and holding two small clay *ghatalu* in his hands. Behind him comes the Elephant Ratham filled with men who distribute new saris along the way, until the supply runs out. Then the Anjali Ratham, but with seven men dressed in saris (not five, as we had been told) sitting atop the vehicle with many others in daily garb hanging from the sides. Then the crowd of Besta men, their fish-net umbrella with its fish decorations held high in their midst. Then the Flow of Milk followed by the Sirimanu Ratham itself.

As soon as the Sirimanu comes into view, a rain of bananas begins, tossed with accuracy and abandon especially at the Sirimanu but also at everyone else in the procession. Thousands of bananas are hurled, from balconies, rooftops, pavement. Good fortune is thought to accrue especially by hitting Bairagi Nayudu.[105] Why bananas, into this watery space,

---

104 Bairagi Nayudu feels the Goddess distributing blessings through him as he is carried by her in the Sirimanu.

105 The news of this hurling of bananas at a goddess tree was what first enticed us to see Paiditalli in Vizianagaram.

25. Bairagi Nayudu Aloft (Boaz Amichay).

26. Bairagi Nayudu Aloft (David Shulman).

27. Sirimanu Underway: Bairagi Nayudu Above it All (David Shulman).

this fluid depth? The simplest exegesis is that the banana is a usual ritual offering, one associated with fertility; thus these bananas hurled with gusto are offerings to the Goddess, just as packets of *pasupu* are thrown at the façade of the Square Temple on the night of Tolellu, and just as a banana bunch is tied to Tree-Paiditalli just below the seat. This is Bairagi Nayudu's explanation. Yet banana reproduction shares an important characteristic with rice—the banana also reproduces through parthenogenesis or, more accurately, through parthenocarpy, the development of the ovary of a flower into a fruit without fertilization. The banana, like rice, is self-reproducing, pollinated by its male flowers. We have seen that the developmental cycle of Paiditalli is one of self-reproduction through phases of maturation. Paiditalli's maturation and harvesting of kingship resonate powerfully with the harvesting of both rice and banana, and indeed with her self-reproducing cosmos.[106]

There is another potential wrinkle here. When she saw a video of the rain of bananas hurled in huge numbers at the Sirimanu, our colleague, the late Galina Lindquist of Stockholm University, exclaimed, 'They're fish!' The watery depths of Paiditalli's cosmos fill with mature fruits that model the maturation of Goddess and kingship, fruits that appear as inhabitants of these depths and that are integrally related to the fishermen who play such a significant role in the biography of Paiditalli—who, like fish, are denizens of the depths of wilderness.[107] Along the route from the Square Temple, the palace of Paiditalli, to the royal Kota, space is, in a sense, flooded by a great wave. This flooding wave, cresting before the royal palace, is also the dynamic of the spectators and worshipers as they watch and hurl bananas.

Paiditalli and her entourage make three journeys from the Square Temple to the Kota-palace and back; but the first is the climactic one, carried high on the surging waves of the crowd's emotion. The Sirimanu passes the tomb of the Sufi Pir situated in the middle of the street and the Raja's flag on the wall opposite, where the royal family should be seated, reaching the main gates of the Kota-palace. In the past, Paiditalli is said to have dipped three times at this spot, lowering the priest towards the entrance of the Kota. Here the procession turns, going back to its starting point, then repeating this route twice more. During each successive

106 Perhaps it is worth noting that the pod of the Tamarind tree resembles the banana in shape.

107 According to Alexander von Rospatt, in Newar rituals of aging in the Kathmandu Valley the shape of the banana is made to represent a fish.

round there is less overt excitement, the high waves of emotion subsidizing into a plenitude, its flow becoming gentler and gentler. Yet there is no lessening of enthusiasm; rather, one senses an increasing fullness and the quietude this brings, a feeling of satiation as a difficult, lengthy journey nears its completion. At the end of the third return, Tree-Paiditalli dips, lowering Bairagi Nayudu to the ground.[108] This once more is a moment of excitement, the return of the priest from the depths of possession by the goddess to her temple. As he alights from the seat, everyone wants to touch him, seeing him as the Goddess, seeking his blessing.[109]

As he is carried aloft, wearing the raiment of a raja, respectfully barefoot in the presence of the goddess, Bairagi Nayudu is fully himself, the priest of Paiditalli, yet formed to receive kingship. We can say that as he is carried, the goddess enters him, possesses him. Yet this is a particularly human perspective.[110] From Paiditalli's perspective—if we may be allowed the hubris of extrapolating her perspective—she absorbs him fully into her interiority, into her cosmos, so that he becomes part of her, part of the wilderness that extends itself through dynamics of similarity becoming difference. This is the significance of the new saris that tie him into his seat. As he is tied, he is held by her, within her. The saris are her, not him. In a way, the saris dress him, so that he is enclosed by her, held next to her, as a mother would carry an infant in front of her. Tied closely to her by the saris, he is, as it were, beginning to be tied into her, absorbed into her, becoming part of her, becoming part of the entirety of her cosmos. The 'small unconscious' he feels (as he once put it to us in English) is just this—internalized by her, he becomes a small part of her entirety, of the perception of a wholeness of cosmos and of himself as perceived from within it.

From this perspective, the world of Vizianagaram is an exteriorization from within the cosmos during this period when Paiditalli comes closest to exteriorizing herself in this way. And it is within herself that the king sprouts into the priest becoming the raja, the priest who *is* the raja, just as the raja is no less the brother of Paidimamba, the younger sister who became a goddess. The priest-turned-king sprouts from within the interior

[108] In 2002, as Bairagi Nayudu was lowered, the rope fastening the high end of Paiditalli to the Ratham slipped, the bottom suddenly rising, and he plummeted half the distance to the ground. Unfazed, he was raised again and lowered once more, slowly.

[109] In 2002 this included the policemen on the spot whom the police superintendent beat off with her lathi.

[110] He says that both Paiditalli and Kanaka Durga are within him as he is carried by Tree-Paiditalli.

of the goddess as he is brought to his palace-fort, where he comes under the protection of the palace Śakti. In this sense, the self-creating Goddess brings the king out of herself *into her own exterior*, into *an extension of herself* that is still herself and, within this, into the kingdom of Vizianagaram that she has sown and grown with her blessings. In a profound sense, within herself she gives birth to the king, her brother—or to her brother, the king. Put otherwise, the king slips out from the Goddess just as Potu Raju emerges from his sister. Older sisters both, younger brothers both. The homologies between sister and brother resonate throughout this cosmos, crucial to kinship, crucial to kingship. The Sirimanu is the apex of a renewed involution of creative wilderness powers within the hardened self-protectiveness of civilization; and of the renewed intimacy between sister and brother that lies at the heart of Vizianagaram kingship, and that likely has broad cultural cachet in Andhra Pradesh.

With each round of Tree-Paiditalli's journey to the Kota, the waves of excitement subside without flattening; rather, as noted, there is the feeling of fullness in the world, of its momentary repletion, of the gentle lapping of waves of deep, imploded feeling. As the sun sets with the third return of Tree-Paiditalli to the Temple, the Sirimanu ends. The pavement of her route has turned slippery and yellowish with the mashed remains of thousands of bananas.[111]

The Sirimanu Ratham stands outside the Square Temple for a week and then is dismantled and stored. Tree-Paiditalli is taken to the Goddess Place close to Bairagi Nayudu's home. Later the tree will be cut into small pieces by Balaji & Co., a firm of tanners that has the right to sell these sections, disseminating her surface traces into a multitude of homes. The *irisimanu* pole becomes Balaji property, and as noted will become the central support of next year's Anjali Ratham. After Paiditalli leaves the tamarind, its wood is thought of more as a husk, a hard cover, a discarded skin, mainly surface with little depth.

Late Tuesday evening after the Jatra, the brahmin priests of the Square Temple and the pot carriers gather at the *saduru*-platform across the street from the temple. Bairagi Nayudu is not among them. Without the rituals of the previous night's Tolellu procession, but accompanied by drummers, they carry the three-piece, Paiditalli~Potu Raju, and eight

[111] In earlier times the Sirimanu began later and journeyed into the night. Police concern for power lines and the possibility of accident during the darkness moved back the starting time.

pots inside the main gates of the Kota. Wrapped in white cloth, little Paiditalli~Potu Raju leads the procession. They go no farther than the main gates, one of the participants says, because this is where Bairagi Nayudu saw Kanaka Durga on the night of Tolellu. Kingship and kingdom are now under the protection of the resident goddess, the Kota Śakti with her affinities to Kanaka Durga. Just inside the entrance is a temporary altar on which the two clay pots of Pot/Mud-Paiditalli already await Paiditalli~Potu Raju. The cloth removed, Paiditalli~Potu Raju is placed between the clay pots and red and black bangles are slipped onto the two side pieces, though not together on the same side piece. Paiditalli is slipping toward childhood and toward the non-differentiation of the fluid deep within herself. This is Paiditalli~Potu Raju's first meeting with the clay pots, the transitional form, Pot-Paiditalli, between Mud-Paiditalli (whose form has dissolved) and Tree-Paiditalli. During Tolellu the two clay pots sat on the swing. Now, in the Kota, Paiditalli~Potu Raju is bangled, red and black, as are infants to protect them against the evil eye and other deleterious forces.

After a lengthy *puja* conducted by the brahmin temple priests, they and the pot-carriers depart with all ten pots and Paiditalli~Potu Raju again wrapped in white cloth. Still at the Kota, Paiditalli offers a meal to Kanaka Durga which no one should observe, during which Paiditalli asks Kanaka Durga to protect the city until she returns. Paiditalli~Potu Raju is placed in the small shrine of Paiditalli, shaded by the three-species tree in the back of the Square Temple. There she joins the many others cut from previous Tree-Paiditallis, a kaleidoscopic enduring return and rejoining of Paiditalli to her presence in her originary shrine, the three-species tree, and of her first differentiation in the human part of her cosmos, between herself and her younger brother, Potu Raju. Six of the pots re-enter the temple; the two clay pots and the two five-metal pots are put on the *saduru*-platform. Later that night the two five-metal pots accompanied by two drummers are carried with soft drumming but without fanfare to the Wilderness Temple. Paiditalli's re-submergence into her own depths continues, culminating two weeks later, when the two clay pots of Mud-Paiditalli are swung away and return to Big Lake.

While the brahmin temple priests are in the Kota performing their puja, Bairagi Nayudu leads a young male buffalo into the empty Square Temple and within the *garbha gṛha*, before Paiditalli. The buffalo had been bought a year before and raised as an offering for this moment. Nayudu releases the rope around the buffalo's neck and this is seized by Scheduled Caste (Dalit) men who have been waiting. They exchange the buffalo for a male sheep which they sacrifice to Paiditalli and eat. A few decades earlier, the buffalo would have been sacrificed to Paiditalli by her priest.[112] Now, without doubt, the intent is present.

11. Uyyala Kambālu—*Swinging Away to the Wilderness*

These two weeks are felt to be a spooky period in Vizianagaram, an uncertain period of betwixt and between, a post-harvest lull, perhaps a time of partial cosmic dissipation. Paiditalli's presence is stretched between the Square Temple and the Wilderness Temple, with direction and momentum shifting towards the latter. Qualities or parts of her have left, others are present. Parts of her are again deep within herself, perhaps resting in the Wilderness Temple, though the goddesses are around, in the city—Kota Śakti in the Palace-Fort; Paiditalli in her Square Temple shrines; Kanaka Durga perhaps diffusely present—but their dynamics generally are shifting toward less activity and hardly any overt interactivity amongst them. Once Paiditalli is swung away to the wilderness, she does not leave the Square Temple from mid-October to mid-May, though every Tuesday she is processed around the temple precincts (in other words, around her cosmos).

All post-Sirimanu rituals are intended to quiet Paiditalli, to soothe her, to please her, to make her softly sleepy in her now more matronly presence. The rite of *Uyyala Kambālu*, of Swinging Away, is especially prominent for the low castes (in particular their womenfolk) who were responsible, mythically, for bringing Paidimamba into the world as the Goddess and who are her practical servants in numerous matters of ritual. The atmosphere within the Square Temple and around the swing is very relaxed. Many more persons are going to visit Paiditalli inside the temple than are gathering around the swing. Among those close to the swing are a group of women of the Bhogumvallu caste who in the past would have been devadasi temple courtesans. Today they are prostitutes.[113] The police presence is minimal (four senior policemen sitting in front of a shop to the side of the swing, and one policewoman).

---

112 Further south, in the region of Rajamundry, such a form of sacrifice to Paiditalli would have been a form of divination, the blood in a plate placed before Paiditalli, the temple gates locked, and on the morrow the patterns of coagulated blood read as portents. The sacrifice was that of Mahisasura, and marked the end of the period of the presence of the Goddess in the community.

113 These women are always auspicious because they never marry and so are never widowed.

We noted that in the Swinging Away the swing itself is thought of as a cradle, the Goddess on the swing cradled within the sacrifice, that is, in the sacrificial skins in which the Goddess was cradled in the past. Indeed, in the past the Swinging Away was even more a rite for untouchables—of Malas (skinners, tanners) and Madigas (agricultural laborers) both of whom performed the blood sacrifices—and of women who had reached puberty. The Malas and Madigas made the swing and swung the seated goddess. According to one informant, in the past the Swinging Away was more explicitly a *pavaḷimpu seva* ritual, that is, one of putting the goddess to bed as is done every evening in her temples. Then a little bed together with the clay pots would be placed on the seat, atop the sacrificial skins. These untouchable castes, like the fishers, belong to wilderness, as do their blood sacrifices and the skins that covered the goddess seat of the swing. The goddess at rest is involuting, contracting, condensing; the more explicit *pavaḷimpu seva* of the past showed that in her resting she became absent from the human world. The swing, then, though it stands adjacent to the Square Temple, is already in wilderness. The likelihood is that Paiditalli is swung away through her two clay pots because they (unlike the metal pots) are transient and soon will dissipate within the fluidity of Big Lake, just as Ammavaru is self-submerging into her own fluid depths.

Thus the swing is yet another *ratham*, a *fluid, modulating version* of the Sirimanu Ratham—as noted, its old wooden posts are made from a Sirimanu ratham after it had completed the Sirimanu; the rope swung between them is taken from the rope used to pull a Sirimanu Ratham. It is thus a kind of *ratham* that, instead of opening depth, encloses and cradles space, or smoothes it, persuading the goddess to leave the space she has opened and encouraging her to involute. Cradled, she returns to her youth, perhaps to her infancy, for there she is sheer potentiality. The goddess is cradled and swung in order to be soothed, relaxed, and sent away to the Wilderness Temple where she can rest, say Bairagi Nayudu and others. He speaks of Paiditalli now as a young girl, and of the swinging as a lullaby.[114] As she re-enters herself, she leaves the traces of her maturation and harvesting on the various surfaces of the kingdom, returning to

---

114 Gell (1980: 221) argues that, 'In all swinging there is an element of self-surrender to a loss of individual equilibrium . . . The swing is an artefact whose use is the modification of mental states.' Swinging (being on the swing) loosens Paiditalli from within, towards childhood, perhaps infancy, as the Sirimanu ratam, from moving forward, curls, coils, into itself, as does Paiditalli within the pots on the swing.

28. Departing Pots on the Swing (Boaz Amichay).

the fruitful depths that human beings will try to activate again in another six months.

Paiditalli's externalization of herself is wholly from within herself, emerging phase by phase, always herself yet different; just as her internalization of herself is wholly into herself. Externalization is maturation, active engagement with and on behalf of the human world within her; internalization is the return to her own primordial being. Her dynamics are those of her cosmos with which we began this discussion—her cosmos opening into itself and closing within itself, its waters flowing from the high depths and the low depths, its mountains fitting into the lakes, its rivers filling the mountains, all of these flowing through and revolving within one another. All of this transpires within, and is externalized within, golden Paiditalli, a watery goddess whose goldenness is fluidity.

Also on the swing is the old tribal, sacrificial axe that was used to make the first gentle cuts in Tree-Paiditalli during the Tevadam rite. Present on the swing are qualities of the phases of Paiditalli's maturation on the human surface of her cosmos: The clay pots absorbed her presence from the mud of Big Lake; the axe helped to bring her as the tree to the city; and the Sirimanu helped bring the king to his kingdom. Qualities that participated in her development now take part in her departure and involution. The first parts or qualities of Paiditalli to be swung away are those of Pot/Mud-Paiditalli, the Goddess born again from the lake. The two clay pots are placed on the swing together with offerings. The pots sit there for some time as the presence of the women becomes more frenzied, with possession by the goddess here and there. The pots and axe are then removed. The swing is cleaned and prepared with fresh offerings for Paiditalli of the Square Temple.

Bairagi Nayudu emerges from the temple holding the small golden *utsava mūrti* of Paiditalli that usually stands in the *garbha gṛiha* in front of her permanent *mūrti*. Also emerging are the axe and the wicker basket containing *pasupu* and *kunkum*. All are placed on the swing. Paiditalli is dressed in a tiny red sari, garlanded, seated on a new women's sari. When Bairagi Nayudu sits on the swing, he holds Paiditalli on his lap. This moment is the final Swinging Away to the wilderness. The *utsava mūrti is* Paiditalli of the Square Temple. Some of her basic qualities are thus now joined to those of the five-metal pots, of Tree-Paiditalli and Pot/Mud-Paiditalli, returning together to the wilderness, to the space of self distant from human being; distant, too, from the depths the goddess has opened, sowed, and reaped, these depths that are kingdom and kingship rejuvenated.

12. *Concluding Thoughts*

Vizianagaram developed as a little kingdom as part of a broad spectrum of a distinctive political culture with particular state-formation, stretching south from the coast of Orissa, including the kingdom of Puri (North Kalinga), through the hill country on the Orissa—Madhya Pradesh—Andhra Pradesh border (taking in the little kingdoms and jungle kingdoms of Jeypore, Bastar, Madugula, Bobbili, and others), to the Godavari River (South Kalinga). In terms of three kingdoms in this spectrum for which we have some information on the annual rituals of the renewal of kingship (Puri, Jeypore, Bastar), we are able to say something further about the relationship of the Sirimanu Jatra to Vizianagaram kingship. This has more to do with what the Sirimanu Jatra is *not*, than with what it *is*. Moreover, we can only learn about this from kingship ritual in these other kingdoms. There we discuss very briefly materials from Puri and Bastar, with a citation of Jeypore.

Immediately evident is that rites of kingship in India generally acted directly on the body and being of the king. If there was a substitute for the king, then the former had to be prepared properly within the palace precincts. When a new king was installed, the tutelary deity of the royal family was called upon to bless the installation (Mayer 1991: 770). The new ruler ascended to the throne, the seat (*gaddi*). At the outset the gaddi had been filled with deity through 'a rite similar to that by which a material object is turned into an image [murti]' (Mayer 1985: 208). In northern Indian kingdoms the gaddi sometimes was described as *the* State (Mayer 1985: 217), for within the gaddi (or within its cushion, which supported the king) were the goddess or goddesses who protected the State with their sakti (Mayer 1991: 774) and who were considered vital to the viability of the king's reign. Mayer (1985: 217) writes that, 'At his installation, then, the ruler... was then joined to it [the gaddi]—some said as a son to a mother (sitting in the mother's lap was the picture given), others as husband to wife.' Galey (1989: 175), writing of the Himalayan former kingdom of Tehri-Garhwal, says that the king 'is actually married to his kingdom and made protector of the earth.' In some kingdoms the king himself worshiped the throne on Navartri/Dasara, the great time of Durga or her homologues.

As we commented, in Indian rituals of the renewal of kingship the focus is the body and person of the king. In the Kalinga region, where there is evidence of annual rejuvenation of kingship, the king either takes an active role or is the recipient of vectors of energy directed at him. In the former kingdom of Puri in North Kalinga, the tutelary deities of kingship—Jagannatha (a form of Visnu), his brother and sister—annually fall ill during the hot, dry season when diseases in the past were rampant. Every twelfth year the deities die and are reborn. Implicitly, they are made ill by smallpox brought by the pox goddess, Sitala (Apffel-Marglin 2008: 142–143). The king, the living incarnation of Jagannatha, has a major role in the annual healing and rejuvenating of the deities. In doing so he becomes an untouchable sweeper (through sweeping and cleansing the chariots of the deities) (Apffel-Marglin 2008: 41, 143). One can say that *as* the king heals the deities he no less actively participates in his own rejuvenation (since he is the reincarnation of Visnu, of Jagannatha).

In the former kingdom of Bastar, for the duration of Navaratri/Dasara the king abdicated his secular power to his prime minister and became an ascetic renouncer, a non-person who performed austerities, who could not be greeted, who slept on the floor, who did not wear shoes nor eat meat. He was substituted for by a stand-in who performed austerities for him. The stand-in sat on a heap of ashes in a pit dug in the durbar hall. At the other end of the pit were a pot of water, a heap of grain, and a sword. The stand-in Raja fasted for the duration of Navaratri, and was not permitted to move his body (Gell 1997: 438). Gell (1997: 440) writes that the stand-in Raja is dead, seated in a grave together with the means of renewal and agriculture: the sword, a ploughshare and sickle; the seed, seed; the water, rain; and the ashes, fertilizer. With the close of Navaratri, the king returns, alive, rejuvenated, and the sacrificed stand-in departs quickly without the two laying eyes on one another. The intimations of the ritual death and rebirth of the king and kingship are clear, while the king is identified with the regeneration of the agricultural year (Gell 1997: 440–441). A quite similar forming of the death and rebirth of king and kingship was reported for the neighboring kingdom of Jeypore (Crooke 1915: 33–34).

In all of these rejuvenations of kingship during Navaratri, the king was either an active participant or was acted upon vigorously and formatively. None of this is so in the Sirimanu complex. The king and then the would-have-been king were expected to touch this ritual cycle at certain points—a few gentle cuts with the sacrificial axe to Tree Paiditalli, giving silk garments to the priest to wear on the Sirimanu, taking a few pulls at the Sirimanu Ratham—yet apparently little more than this. Given that Vizianagaram kingship likely shared qualities of transformative renewal with these other little and jungle kingdoms, where was this most likely to have taken place in Vizianagaram? Most probably inside the Kota, inside the palace quarters, inside the Round Mahal with its Durbar Hall and throne, in the presence of the immoveable Kota Sakti, the energizer and protectress of kingship within itself. In there, it was likely that the king annually went through some form of death and rebirth, perhaps during Navaratri. Unfortunately the Royal archives of Vizianagaram are either destroyed or lost, so our supposition cannot be confirmed. Nonetheless, continuing with our inference puts Paiditalli into a different focus.

What is the significance of the presence of the Sirimanu in Vizianagaram? The immediate response is historical. Vizianagaram's military power finally was decimated at the end of the eighteenth century by its erstwhile ally and partial controller, Britain, at the battle of Padmanabha. After this the kingdom entirely abandoned military ventures and concentrated on political intrigues in Madras and on its own developing, vibrant cultural life which reached its apex in the second half of the nineteenth century. Vizianagaram kingship was a major force in this inward turning and in encouraging this South Indian cultural renaissance. Though we are unsure of the exact dates of its initial practice, the Sirimanu may well be part of this creative cultural movement or received impetus from this in the nineteenth century.

We could rephrase the above in the following way. The Sirimanu complex is the people's rites of kingdom and kingship and their rejuvenation. They were the people's creative response to the crisis of kingship in Vizianagaram. This response entailed the cosmic perceptions of a goddess whose presence was prevalent in rural areas of the region and who was re-originated precisely as the people's protectress of Vizianagaram kingdom and kingship in the face of the crisis of Bobilli. Note that at no point do the Paiditalli rites penetrate the palace interiors of the Kota where the Kota Sakti reigns and where the king's rites of kingship likely were done. Neither do the Paiditalli rites clash at any point with where kingship rites were prominent nor do the Goddess rites belong to a particular sector of the kingdom (as could be said for Bastar, where in the last phase of the Dasara rites of the death and rebirth of the king, he is 'kidnapped' by 'tribal' peoples and re-formed as Dantesvari, the goddess of land and people, as whom he is returned to his palace [Gell 1997: 439, 442]).

In the people's rites of Paiditalli it is the self-sacrifice of the Goddess for the sake of kingdom and kingship that always is at the forefront; and this is tied directly to the historical event of her brother's, Raja Rama Raju's, death at Bobilli, and not to the death and re-birth of king and kingship as such (as was the case in Bastar, and perhaps in Jeypore and Puri as well). The Sirimanu unusually and ironically *celebrates* the tragic death of the king, and the focus in the Sirimanu complex is on the king's sister and her deep tie to her brother. All of this feels like rites of the people in a cosmos with which, in terms of cultural common sense and its truths, they are intimate. In a *vital* sense, the rites of Paiditalli entail and practice her existential and phenomenal existence and not those of kingship and king, though in her cosmos she indeed takes responsibility for them. It is worth emphasizing again that Paiditalli throughout is a relatively pacific goddess devoted to self-sacrifice for the sake of the growth of kingdom and kingship, and this in a previously self-perceived warrior kingdom.

By calling the Sirimanu complex the people's rites of kingdom and kingship we are suggesting that the likely impetus for these were neither directly in royalty nor in the usual state priesthood of brahmins (who were not significant in Vizianagaram kingship outside of the palace). Perhaps the impetus (if not the initiative) came from goddess priests within the atmosphere of a kingdom turning inward and perhaps searching inward and discovering forms of shaping the world other than those of war, culminating in the cultural effervescence of the nineteenth century.

Our argument has implications for just how the Sirimanu rejuvenates kingship. Kingship—perhaps we should say, maleness—enters the

Sirimanu through the appearance of the three-piece, the creation and joining of Paiditalli~Potu Raju, at Tolellu, the apex of the heightening and activation of the Goddess' fertility. The three-piece is cut from, born from, Tree-Paiditalli in her cosmos. Potu Raju in this assemblage emerges from his older sister. Maleness emerges from femaleness. It is here that Paiditalli creates gender. But in this regard the identities of Paiditalli and Potu Raju are not (quite) distinct. In the three-piece, sometimes femaleness is predominant, sometimes maleness, but both the Goddess and her younger brother share these qualities, and maleness remains integral to femaleness. Their beings in this cosmos are permeable, their qualities fluid and flowing into one another, to borrow loosely from the thinking of McKim Marriott (1989, see also Busby 1997, Daniels 1984).

With Potu Raju, the Buffalo King, kingship is born within Paiditalli's cosmos, with herself as its protectress, the guardian of her younger brother, the king. King Rama Raju was the older brother of Paidimamba, the ancestress of Paiditalli, and lost his life at Bobilli because he wouldn't hearken to his sister's advice. Now kingship depends on the power, yet no less on the wisdom, of Paiditalli. Her wisdom may be why Paiditalli *insists* on appearing as the Sirimanu tree, re-creating, giving birth to, her now younger brother, the Buffalo king. The identity between Potu Raju and the kingship of Vizianagaram (including its incumbent) is precisely here—Potu Raju and the king are one. Royalty in the Sirimanu is a family composed of sisters and brothers who together fully regenerate the kingdom as Tree-Paiditalli deposits the king (perhaps an infant?) at the doorstep of his palace-fort, the domain of Kota Śakti.[115] It is the people's rites of kingship in Vizianagaram that catch the significance of the sister-brother tie in Andhra Pradesh more generally and make this a fulcrum for the regeneration of kingdom and kingship. At the top of Tree-Paiditalli in the Sirimanu, in the figure of her priest, Bairagi Nayudu, there is the extreme condensation of Paiditalli's biography. The originary Paiditalli appears as a stick held onto by Bairagi Nayudu (for dear life, as he once told us). Attached to this stick is the mat on which Paiditalli sat beneath the three-species tree (which may well be her as well), her originary

[115] Oral traditions in India do not obviate conundra of incestuous unions, which may be cleaned up in more canonic or priestly versions. For example, Galey (1989: 175), writing on the Himalayan former kingdom of Terhi-Garhwal, states that, '... much is still unexplained in the incestuous form of union developed by the oral tradition where the king marries the earth which is also his sister, his mother, and his daughter.'

shrine in Vizianagaram after she first emerged from Big Lake. Wrapped in silk cloth is the three-piece, Paiditalli~Potu Raju/incumbent king, all three pieces cut from Tree-Paiditalli and resting on Bairagi Nayudu's lap, just as his ancestor the chief-minister held Paidimamba on his lap prior to her brother's death.[116] And perched on the Sirimanu is Kanaka Durga, the protectress of Paidimamba, who enabled the latter's transformation into Paiditalli.

The sequence of forms Paiditalli takes during her ritual period in Vizianagram tells us more of the energies of regeneration activated here. Paiditalli emerges from Big Lake as formless, almost amorphous ooze. Mud-Paiditalli then enters the two clay pots specially made for her. Paiditalli acquires firm shape. The interior of Pot-Paiditalli is circular, perhaps spiral, her trajectories of energy going round and round, coiling within and through themselves through interior dynamics, intensifying more and more. The pot is homologous with the womb, and within herself Pot-Paiditalli is developing the differentiation of her cosmos, which will appear at Tolellu as gender distinction. With the appearance of Tree-Paiditalli, she becomes straighter, her trajectory potentially more linear, one end pointing downward into the earth, the other, potentially upward, forward, into the heavens. With the appearance of the three-piece, Paiditalli~Potu Raju, during Tolellu, gender comes into existence, though female and male qualities are blurred fluidly into one another (just as in Mud-Paiditalli her different qualities were quite blurred and fluid amongst themselves). With the appearance of the infant-like three-piece of Paiditali~Potu Raju and the doing of *Sirilu* (the First Furrow of generating fertility), Tree-Paiditalli should be understood as having both female and male qualities.

The following morning, Tree-Paiditalli on the Sirimanu Ratham, elevated to a 45 degree angle, looks more male-like, perhaps a female-male, thrusting forward, opening space, the throngs of onlookers sighing and moaning her ejaculation, tailing off (especially at the close of first arrival at the Kota entrance) as she lowers herself to permit the king to dismount. On the Sirimanu Ratham all forms are linear—the rectangular mat, the stick, the three-piece. All pots, all dynamics of coiling and intensifying,

116 The three-piece may resonate with the three-species tree. Here we will not enter into the possible significance of threes or triads in the entire Paiditalli complex, though without doubt this is important—the three goddesses (Paiditalli, Kota Sakti, Kanaka Durga), the three-species tree, the three-piece, the three rounds of the Sirimanu between temple and fort-palace, and likely more.

have disappeared. But on the evening after the Sirimanu, pots reappear as the two clay pots meet the three-piece on a temporary altar inside the main entrance to the Kota. After this it is linearity that disappears as the three-piece enters the small shrine adjacent to the three-species tree and the pots that return, as Paiditalli begins to intensify her involution, returning deeper within herself, in a sense deeper within her own womb, until the swinging away of the two clay pots on the swing that is thought of as a cradle.

A Tantric reading of Paiditalli's rhythms might see a progression from femaleness to maleness to the uncoiling of energy. The pot is not only a womb; it also resembles the cakra centers of subtle energy in Tantric Yogic physiology. The female energy of the Goddess coiling and intensifying in the pot is not dissimilar to that which occurs through meditation and ritual in the Yogic cakras, especially that at the base of the spine, as the sleeping Kundalini is shaken awake. In Tree-Paiditalli this coiling uncoils and straightens. In Tree-Paiditalli on the Sirimanu Ratham traveling to the Kota, especially in the first journey, uncoiling energy, unleashed, rises as would the Kundalini to the top of the skull, to the top of the Tree and beyond, thrusting forward with male energy. Tree-Paiditalli on her Ratham is the synergistic blending of female and male energies of the Goddess and Potu Raju that together re-energize, give birth to, the infant king deposited at the entrance to his palace-fort. And what then of the banana, hurled by the thousands at the Sirimanu Ratham—the simple banana that first brought us to the Sirimanu Jatra and set us on our journey of exploration into a people's rites of rejuvenating kingdom and kingship? Of course one could always say, trivially, that the banana has a phallic shape. More important in our eyes is the self-pollinating and self-generating nature of the banana plant; and we would want to add that the Sirimanu Tree, blending in its intensities both female and male energies, may well be described as such a self-generating phallus emerging from the female pot-wombs of Paiditalli. Perhaps there is, indeed, an implicit Tantric logic in Vizianagaram—for what is Tantra if not the formalization of just such local conceptual logics as we find in the Paiditalli rituals? We know that Tantric Yogis had a major role in the life of the Vizianagaram court alongside the wrestlers and tiger dancers and poets. At the least we can say that showering the Sirimanu Ratham with bananas is a tremendous intensification of the dynamics that have gone into this people's shaping, that is, rejuvenating, their kingdom and their king.

We would like to stress again, in conclusion, the perhaps unexpected notion that goddesses, like other nourishing and living beings, themselves need to be grown, nurtured, fermented, and matured, and that it is the work of human beings to see them through these processes year after year. Paiditalli is a living goddess, hence never a given, a datum, a "fact." Like all other living beings, she exists in varying intensities and uneven textures; it would be ludicrous to claim that she is always a single set of attributes or features, or that her self-awareness is steady or fixed, or even that she is uniquely female—when, as we have seen repeatedly, the very division of femaleness and maleness is a categorical feature of the surface. On another level, as a goddess endowed with depth, indeed defined by her depth, Paiditalli is capable of generating maleness in the form of her intimate male self, her brother and, at the next remove, a king. We have attempted to see, with her devotees, into the deeper innerness of this goddess and to watch her emerge from it, or through it, in order to draw close to her people, only to depart with their help back into that deep interior when this year's task is done.

Bibliography

Akundy, Anand, n.d. 'Bards and goddess festivals: The contribution of Pombalas to the Gangamma jatra of Tirupati, South India.' Ms.

Apffel-Marglin, Frederique, 1981. "Kings and Wives: The Separation of Status and Royal Power." *Contributions to Indian Sociology*, 15, no. 1–2: 155–181.

——, 2008. *Rhythms of Life: Enacting the World with the Goddesses of Orissa*. New Delhi: Oxford University Press.

Assad, Maria L., 1999. *Reading with Michel Serres: An Encounter with Time*. Albany: SUNY Press.

Balzani, Marzia, 2003. *Modern Indian Kingship. Tradition, Legitimacy & Power in Rajasthan*. Oxford: James Currey.

Berkemer, G., 1993. *Little Kingdoms in Kalinga: Ideologie, Legitimation, und Politik Regionaler Eliten*. Stuttgart: Franz Steiner Verlag.

Biardeau, Madeleine, 2004. *Stories About Posts: Vedic Variations Around the Hindu Goddess*. Chicago: University of Chicago Press.

Busby, Cecilia, 1997. "Permeable and Partible Persons: A Comparative Analysis of Gender and Body in South India and Melanesia." *Journal of the Royal Anthropological Institute* (N.S.) 3, 261–278.

Caldwell, Sarah, 2001. *Oh Terrifying Mother: Sexuality, Violence and Worship of the Goddess Kali*. New Delhi: Oxford University Press.

Chatterjee, Kamala, 1978. "Brahman as Ignorant." *The Journal of the Indian Academy of Philosophy* 17, no. 2, 1–16.

Crooke, W., 1915. "The Dasahra: An Autumn Festival of the Hindus." *Folklore*, 26, no. 1: 28–59.

Daniels, F. Valentine, 1984. *Fluid Signs: Being a Person the Tamil Way*. Berkeley: University of California Press.

Daryn, Gil, 2006. *Encompassing a Fractal World: The Energetic Female Core in Myth and Everyday Life*. New York: Lexington Books.
Dirks, Nicholas B., 1987. *The Hollow Crown: Ethnohistory of an Indian Kingdom*. Cambridge: Cambridge University Press.
Falk, Nancy E., 1973. "Wilderness and Kingship in Ancient South Asia." *History of Religions*, 13, no. 1: 1–15.
Flueckiger, Joyce, n.d. 'Guises, turmeric and recognition in the Gangamma tradition of Tirupati.' Ms.
Galey, Jean-Claude, 1989. "Reconsidering Kingship in India: An Ethnological Perspective." *History and Anthropology* 4, pt. 1, 123–187.
Gell, Alfred, 1980. "The Gods at Play: Vertigo and Possession in Muria Religion." *Man* (N.S.), 14: 219–248.
——, 1997. "Exalting the King and Obstructing the State: A Political Interpretation of Royal Ritual in Bastar District, Central India." *Journal of the Royal Anthropological Institute*, 3, no. 3: 433–450.
Good, Anthony, 1983. 'A symbolic type and its transformations: The case of South Indian ponkal.' *Contributions to Indian Sociology*, 17: 223–244.
——, 2000. "Congealing Divinity: Time, Worship and Kinship in South Indian Hinduism." *Journal of the Royal Anthropological Institute* (N.S.), 6: 273–292.
Gold, Ann Grodzins, 2003. 'Vanishing: seed's cyclicity.' *Journal of Material Culture*, 8, no. 3: 255–272.
Handelman, Don, 2005. 'Introduction: Why ritual in its own right? How so?' In *Ritual in Its Own Right*, edited by Don Handelman and Galina Lindquist, New York: Berghahn, 1–32.
Handelman, Don and D. Shulman, 2004. *Śiva in the Forest of Pines: An Essay on Sorcery and Self-knowledge*. Delhi: Oxford University Press.
Marriott, McKim, 1989. "Constructing an Indian Ethnosociology." *Contributions to Indian Sociology* (n.s.), 23, no. 1: 1–39.
——, 1998. 'The female family core explored ethnosociologically.' *Contributions to Indian Sociology* (n.s.), 32, no. 2: 279–304.
Mayer, Adrian C., 1985. "The King's Two Thrones." *Man* (N.S.), 20: 205–221.
——, 1991. "Rulership and Divinity: The Case of the Modern Hindu Prince and Beyond." *Modern Asian Studies*, 25, no. 4: 765–790.
Moreno, Manuel, 1992. "Pacamirtam: God's washings as food." In Khare, R. S. (ed.) *The Eternal Food: Gastronomie Ideas, and Experiences of Hindus and Buddhists*. Albany: SUNY Press, 147–177.
Mrokkina varamul' icce callani talli sri sri sri paiditalli ammavari katha caritra. 1994. Vizianagaram: privately published.
Narayana Rao, Velcheru, David Shulman, and Sanjay Subrahmanyam, 2001. *Textures of Time: Writing History in South India, 1500–1800*. Delhi: Permanent Black.
Nidbach, Maayan, 2013. Merkaziyuto shel Musag Hazman Bitfisat Halashon vehahavaya shel Bhartṛhari. M.A. Thesis, Hebrew University.
Nuckolls, Charles W., 1996. *The Cultural Dialectic of Knowledge and Desire*. Madison: University of Wisconsin Press.
Pusupati rajulayokka purvottaram. In Pachauri, S. K. (ed.), *Vijayanagaram jilla kaphiyyatulu*. Hyderabad: Andhra Pradesh Government, n.d., 1–10.
Rama Rao, V. V. B., 1985. *Poosapati: Ananda Gajapati Raju*. Hyderabad: International Telugu Institute.
Roghair, Gene H., 1982. *The Epic of Palnadu*. Oxford: Clarendon Press.
Sarma, A. Vizaya Dattatreya, 1994. *Vizianagaram District in the Lap of History*. Vizianagaram: Mansas Trust.
Schnepel, Burkhard, 1996. "The Hindu King's Authority Reconsidered: Durgā-Pūjā and Dasarā in a South Orissan Jungle Kingdom." In Boholm, Asa (ed.), *Political Ritual*. Goeteborg: Institute for Advanced Studies in Social Anthropology, 126–57.

Schnepel, Burkhard, 2002. *The Jungle Kings: Ethnohistorical Aspects of Politics and Ritual in Orissa.* New Delhi: Manohar.
Shulman, David, 1995. "First Man, Forest Mother: Telugu Humanism in the Age of Krishnadevaraya." In *Syllables of Sky: Studies in South Indian Civilization in honour of Velcheru Narayana Rao*, edited by D. Shulman, Delhi: Oxford University Press, 133–64.
——, "Waking Aja," in Bronner, Yigal; Shulman, D.; and Tubb, Gary, in press.
Sundaram, Lanka, 1946. "Revenue Administration of the Northern Circars (Part 2)," *Journal of the Andhra Historical Research Society* 15: 1–118.
Waghorne, Joanne Punzo, 1994. *The Raja's Magic Clothes: Re-Visioning Kingship and Divinity in England's India.* University Park, PA: Pennsylvania State University Press.

INDEX